Lionheart
a journey of the human spirit

JESSE MARTIN
With Ed Gannon

This paperback edition published in 2023

First published in 2000

Copyright © Jesse Martin and Ed Gannon 2000

All rights reserved. No part of this book may be reproduced or transmitted in any form or by any means, electronic or mechanical, including photocopying, recording or by any information storage and retrieval system, without prior permission in writing from the publisher. The *Australian Copyright Act 1968* (the Act) allows a maximum of one chapter or 10% of this book, whichever is the greater, to be photocopied by any educational institution for its educational purposes provided that the educational institution (or body that administers it) has given a remuneration notice to Copyright Agency Limited (CAL) under the Act.

Published by Jesse Martin and Ed Gannon

Email: info@lionheart.fund
Web: www.lionheart.fund

Lionheart: a journey of the human spirit.
ISBN 979 8 36361 905 2

Text prepared with the assistance of Ed Gannon
Maps and diagrams by Chris Heywood
Typeset by J&M Typesetting
Cover photo by Serge Thomann
Back cover photo by Mark Smith / Herald Sun

 A catalogue record for this book is available from the National Library of Australia

*To Mum, who made it all possible.
Dad, my best friend.
And the dreaming child in all of us.*

Lionheart

Also by Jesse Martin

Kijana: The real story
I do believe her dark clothes were the attraction

THE WHITE HOUSE

WASHINGTON

September 9, 1999

Jesse Martin
Melbourne, Victoria
AUSTRALIA

Dear Jesse:

I recently learned about your effort to become the youngest person to sail around the world single-handedly, and I want to extend my best wishes to you for a successful trip. Such a journey would be an immense challenge for anyone, and for a 17 year old it is especially remarkable. You are an impressive young man, and your courage and determination are an inspirational example for all young people to follow.

I encourage you to continue to set high goals for yourself and to pursue your dreams. You can do anything that your imagination, effort, and talent will let you achieve. Best wishes.

Sincerely,

Bill Clinton

Online Content

This paperback edition is without author images. Follow the link below to access more content online.

Additionally, 'Lionheart The Jesse Martin Story' is a 60 minute program which screened around the world on National Geographic Channel and can be accessed by scanning the below QR code, or by visiting:

https://www.lionheart.fund

Contents

    Map of Jesse's Voyage viii
    Author's note x
    Prologue 1

1 The First Steps 3
2 From Belize on a Breeze 27
3 Making the Dream Come True 60
4 The Mad Rush 84
5 Reality Bites: Australia to New Zealand 114
6 On to Everest: New Zealand to Cape Horn 143
7 Through a Mind Field: Cape Horn to the Azores . . . 176
8 Please, God, Stop this for Me: Azores to Cape of
    Good Hope 203
9 The Final Run: Cape of Good Hope to Australia 245
10 Beyond the Waves 274

    Appendix 1: Equipment List 290
    Appendix 2: Glossary 293
    Appendix 3: Parts of the Boat 296
    Acknowledgements 299

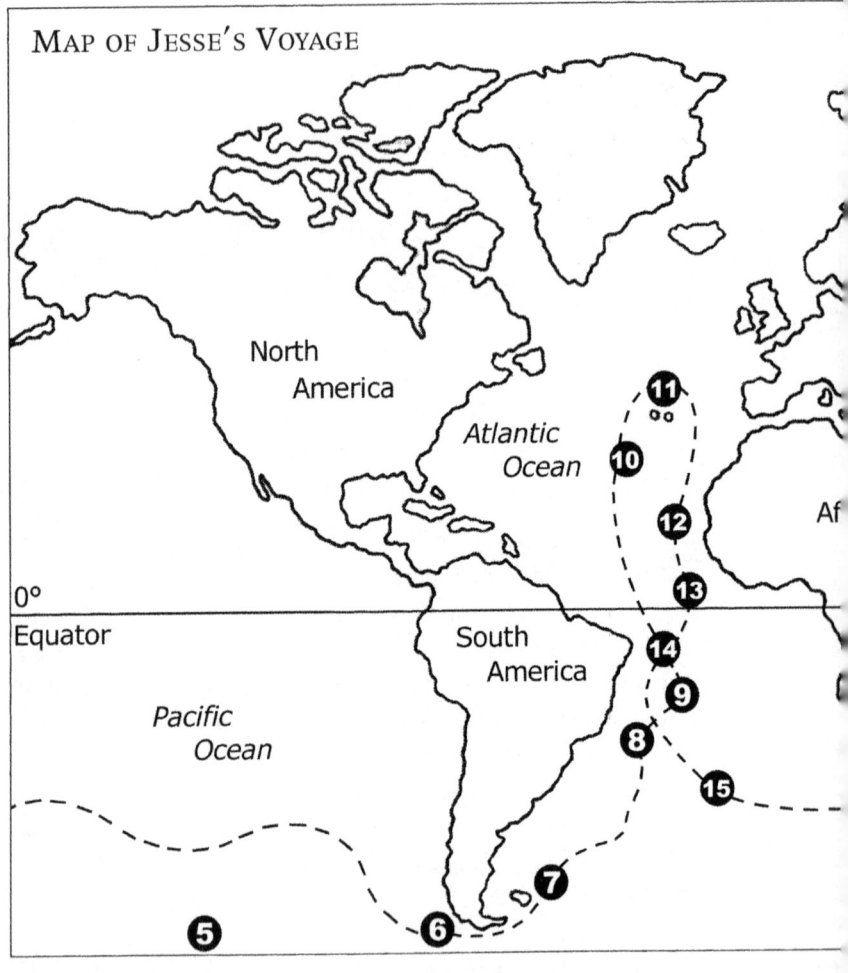

1   departed 7 December 1998,
    arrived 31 October 1999
2   Christmas day
3   met fishermen
4   first bad weather
5   position of rescued Autissier
6   first proper knockdown
7   second knockdown
8   whale encounter
9   becalmed for four days

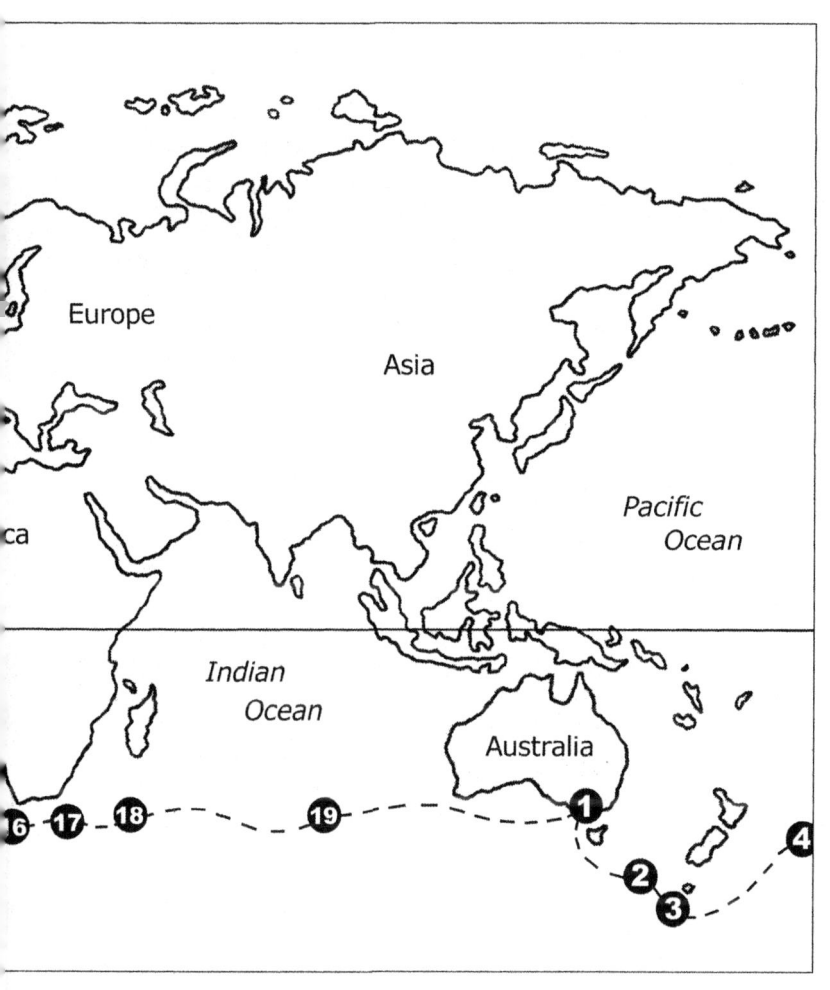

10 close call with tanker
11 met family
12 pirate scare
13 furler problem
14 passed over previous track
15 third knockdown
16 force 10 storm
17 no power
18 eighteenth birthday
19 mid Indian rough patch

Author's note

Throughout the book kilometres and miles are used to refer to distance. Distances on land are recorded in kilometres, while distances at sea are in miles. All reference to miles refers to the metric measurement of nautical miles (1.852 kilometres), which is longer than the land mile of 1.6 kilometres. All speed at sea is in knots, the term for nautical miles per hour. Distance at sea can be worked out by one degree of latitude equalling 60 nautical miles. The boat industry also tosses up some anomalies, referring to the length of craft in imperial measurements, which is why boat lengths are referred to in feet and inches throughout the book. While explanations for sailing terms are given throughout the book, a full glossary can be found on page 245.

Many thanks to John Hill for the use of his technical notes in the writing of this book.

Prologue

> It is better to live one day as a lion than ten years as a lamb.
>
> — Anon

The world is a funny place.

It's hard to imagine, as you stand in your backyard or lie in your bed, that there is a spot on this amazing globe directly opposite you. If you drilled a hole directly through the earth, where would you come out?

At 5.36 p.m. on 7 December 1998, I sailed from the safety of my home waters of Melbourne, Australia, on my 34-foot yacht *Lionheart*, and set course for that other point—latitude 38°18′N, longitude 35°22′W. And when I got there, I kept going, returning to Melbourne on 31 October 1999, 328 days and roughly 27,000 nautical miles later. In my quest, I became the youngest person to circumnavigate the globe solo, non-stop and unassisted.

Why? Simply, for the adventure. I had dreamt of sailing around the world, so that's what I did. If we don't live our dreams, what's the point of living?

Lionheart

Lionheart—the name of the boat that carried me there and back, and also the title of this book—is a story of that journey. More than that, it is a story of the human spirit, the very same spirit that lives in all of us, and what it can achieve when put to the test.

CHAPTER 1

The First Steps

Sunday, December 7, 1998
With the genoa unfurled for the first time, I passed Sorrento, then Portsea, making my way through the South Channel to the starting line—the heads of Port Phillip Bay, the most treacherous port entry and exit in the world.

I cut a thick slice of salami, then rushed up on deck to correct the wandering steering that had once again deviated slightly towards land. Maybe *Lionheart* was trying to tell me something. Maybe she'd prefer to stay at home in the shelter and safety of the bay rather than enter the unknown of Bass Strait. Did she know we were about to take on the world? Moments later I crossed the line, and my new life began.

What makes a seventeen-year-old decide to sail around the world? I'm not exactly sure, I was actually fourteen when I first started to think about doing so. When I sailed from Port

Phillip Bay on 7 December 1998, the trip was the culmination of years of dreaming. Others may have thought I was a foolish young man, but I'd been working towards that dream for a long time.

Why? That's the question I get asked most. And one of the reasons behind this book. I don't just want to tell the story of how I sailed around the world on my own, but to reveal why a teenager would want to leave the comfort of home for eleven months at sea, and what I learnt from the experience.

It has been said that every great adventure begins with one small step. It's clichéd, but it's true. I've taken thousands of steps to become the youngest person to sail solo, non-stop and unassisted around the world.

But what was that first step? Was it sailing through the Port Phillip Heads, my official starting point of the trip? Was it waving goodbye to family and friends at the Sandringham Yacht Club? Was it when my major sponsor agreed to commit $160,000 to my trip? Was it that moment, at fourteen years of age, when I first dreamt of sailing around the world? Was it my previous adventures? Was it the first time I stepped aboard a boat? Was it when I was born?

Who knows, but I suspect Mum and Dad had a fair bit to do with it.

I suppose my story begins in 1979, when my parents, Kon and Louise, did something quite radical for a happily married young couple living a comfortable suburban life in Melbourne. They sold their cars, rented out their house and set off in a Volkswagen beetle to see the world. I was born on that trip, and their spirit of adventure lives on in me today.

Up the east coast of Australia they drove, to Far North Queensland and Darwin, where they ditched their car and made for Bali, spending several months backpacking through

The First Steps

Asia. From there it was on to Europe, arriving at Dad's birthplace, Germany, in June 1980.

They decided to stay in Germany for a while. Dad got a job driving trucks while Mum worked in a supermarket and as a hotel room cleaner in Bonn. But the travel bug bit again so they bought another VW, this time a kombi van, and travelled through the rest of Europe.

A few weeks into this trip Mum discovered she was pregnant. They decided that when the time came for me to arrive, the birth would be as natural as possible. Dad's relatives pointed them in the direction of a small village called Dachau, near Munich, that had quite a reputation for its natural birthing clinics. Of course, Dachau is better known as the site of the Nazis' infamous World War II concentration camp.

I was ten days late entering the world. Luckily, for it was only a few days before I was born that I was given a name. It came to Mum when her frustration at spending her final days of pregnancy in the sweltering heat of a campervan finally boiled over. On this day she lashed out and kicked the van's radio. It jumped from a German station to the American armed forces' radio station. Playing was the song 'Jesse' by Carly Simon. She decided on the spot that if I was a boy, I'd be called Jesse. She'd already decided on Heidi if I was a girl, much to my German relatives' horror, as Heidi was a pretty common name in Germany. Mum thought it would be sufficiently exotic in far-off Australia.

I arrived safely into the world at the local hospital on 26 August 1981, a helpless little thing, according to Mum. And just in time to cause my first controversy, which even made the German newspapers. 'Baby in der Mühle der Bürokratie,' said the headline, which translated as 'Baby in the mill of bureaucracy.'

To get a birth certificate in Germany, Mum and Dad had to present their marriage certificate, which they duly did. They were married in Melbourne in 1975 in the Seventh Day Adventist Church. But the local authorities had never heard of that church, and wanted to check the authenticity of the church and the marriage certificate before they'd issue my birth certificate. Mum and I were stuck in hospital for two weeks until the matter was sorted out. Mum still wonders what would have happened if she had been a single mother. Would she still be in hospital as the authorities demanded a marriage certificate?

As soon as my parents got the birth certificate, it was on to the Australian embassy at Bonn to have me put on Mum's passport so I could begin my life of travel. Mum wanted to come home to Australia, so we made our way to Hungerford's carpark in London, where backpackers—mainly Aussies—bought and sold campervans. Selling our campervan proved difficult initially as it wasn't a pop-up vehicle. But luckily we came across someone planning to travel to a part of Africa where, if you cross a river, you are charged extra for a pop-up van. Strange rule, but it enabled my parents to sell the van.

We had to virtually dash from the carpark to make it on our flight to Australia. We must have looked a sight entering the airport: Mum had me slung across her front and a pack on her back, and Dad had a pack on his back and another on the front.

I was five weeks old when I touched down in Australia. And although I've left her shores many times since, I've always returned.

~

The First Steps

We settled into a bungalow behind Mum's parents' house in Upwey, an outer Melbourne suburb. But Mum and Dad soon grew restless, and wanted to move on, to live life their own way. So they bought an HR Holden station wagon and drove up the east coast looking for some land to build a home. They got to Cairns, 3500 kilometres north of Melbourne, where they were told of a one-hectare block at Cow Bay in the Daintree rainforest 120 kilometres further north. The agent was at least honest. He said they probably wouldn't be interested as it had no electricity, no water, no anything.

He was wrong. It was ideal, as my parents wanted to turn their backs on a boring suburban life. They paid $23,000 for their own piece of tropical paradise. Before we could move there, however, we had to go back to Melbourne while Dad worked to raise some money to pay for the land.

In Melbourne, my mother became pregnant with my brother Beau. Dad decided to move back to the Daintree block so that he could build a shack for us, which meant he was away for a couple of months of Mum's pregnancy. He came back four days before Beau was born, on 6 June 1983.

Five weeks after Beau was born, we loaded up an old Toyota Land Cruiser with a boat on the roof and a caravan behind, and headed north. Beau travelled in a fruit box at Mum's feet. I was just on two years old.

When we first arrived, things were pretty basic. Home was four poles, a roof and a floor—the correct term is a 'humpy'. There were no formal windows and doors, and snakes slithered outside. Sometimes Mum and Dad would return from swimming at the beach or a trip into town to find hoof marks from wild pigs throughout the house. For water there was a creek with a waterfall, so Dad rigged up a system to pump water into the house. Electricity came from a generator.

Lionheart

When I was two-and-a-half years old, my parents realised I'd developed a serious stutter, which worsened over about six months until I was virtually unable to talk. Mum freaked out. After doing some reading, she took me to a child psychologist in Melbourne who confirmed her suspicions. It seems Mum and Dad were going through a bit of a rocky patch at the time, and although I was too young to understand or remember it, I was able to absorb a lot of what was happening, which resulted in the stutter. I was actually feeling the tension between them. The doctor told Mum that the only way to fix things was for them to encourage me in everything I did and to build my self-confidence.

From that moment on, with my family's support, I was made to believe I could do anything. I had no boundaries in the rainforest. I could run anywhere, explore and experience new things with the blessing of my parents. My life was real Jungle Boy stuff. It was a life of fishing and snorkelling on the reef. Mum has photographs of me running proudly about the forest naked. Dad picked up labouring work to keep some food on the table, and sometimes he'd shoot a wild pig, which we'd eat or use for bait.

The freedom continued throughout my childhood and teens. Looking back, those pivotal decisions my parents made to help me overcome my stutter probably set me on the path I would choose, and to achieve what I have.

By 1985 we'd moved again, this time to Sydney, so Dad could get work as a builder to earn some money to build a bigger house on the Daintree block. We ended up buying an old Victorian house in suburban Sydney and renovating it.

In 1987, after two years in Sydney, Dad decided it was time to move back to the Daintree. Mum didn't. So while Dad headed north, Mum, Beau and I moved back to Melbourne.

The First Steps

I was five years old, and didn't see much of my father during the next six years.

We moved into a bungalow behind Pop and Gran's house in Upwey and I soon started school at Upwey South Primary School. I loved it. Beau and I spent a lot of time with Pop in his shed, where he showed me how things worked. If a toy was broken, he'd help us fix it.

'It can't be fixed, Pop,' we'd say.

And he'd reply, 'Of course it can, we'll fix it.'

That simple message taught me a lot. One of my biggest regrets is that Pop did not live to see me undertake my voyage. He died eight months before I set sail from Port Phillip Bay. For him, to see me set off and come home safely would have made him the proudest person in the world.

∼

What I remember most about my primary years were the great holidays that Beau and I could look forward to when school broke up. Mum would usually take us out of school a week before the term holidays and let us go back a week into term—she reckoned there were some things you couldn't learn in the classroom.

Mum took me to the United States when I was five years old. She always made the effort, when Beau and I were young, to take us to places that were a bit unusual, like Darwin, Uluru or even Asia. Like any normal primary school kids, we wanted to go to McDonald's and the movies, but Mum would sit and talk to us about what we could do.

'Boys, we have an option here,' she would say. 'We can do those things, or we can do the interesting stuff.' Luckily, the interesting stuff usually won. The decisions came from Beau and me—Mum never forced them on us.

One trip stands out in my mind. When I was in Grade 5

Mum took Beau and me backpacking across southeast Asia for five weeks. It was amazing. We went to Thailand and visited the Golden Triangle, where we saw opium fields, and bridges built from bamboo over steep ravines. We went on an elephant trek and visited the island of Ko Samui, which was a real buzz. For an eleven-year-old, catching buses crammed with chickens in baskets while people threw up around you was an incredible experience. I fell in love with Thai food on that trip and am forever trying to recreate it at home, without much success.

Beau and I spent most of our time in Thailand scouring the local markets for flick knives and sling shots. We were fascinated by weapons, and had collected quite an arsenal, which we used to catch birds at Upwey.

The following year, I was off overseas again on an exchange for four weeks to Mexico. It sounds strange, but I really didn't think I was doing things differently from the other kids at school. I should have realised I was the only kid bringing flick knives and nunchakus to show and tell.

On my eighth birthday I had my first taste of sailing, when a friend took Mum, Beau and me out on a yacht on the Gippsland Lakes. I can't say it was a pivotal moment in my life, but it was the first time I had sailed, even if the highlight of the weekend was when the yacht ran aground. For the record, I was not at the helm at the time.

I was a good student at primary school, one of those kids who got solid marks, was good at sport and never got into trouble. I actually had an individualistic bent, even then. Although most of my sports were team sports, I'd basically play for myself. If the team won I'd only be happy if I had played well. If I played my best and we lost, I'd still be happy. But I was not a selfish player—I would happily pass the ball if someone had a better chance of scoring. I knew in my mind that I'd

The First Steps

contributed to that goal, and that was satisfying.

I got my first job when I was ten, delivering pamphlets after school and on weekends to five streets in the Upwey area. It was hard work. Some weeks Beau and I would package five pamphlets for each letter box, walking from box to box, while Mum or Pop drove along, carrying our supply of pamphlets. We actually expanded into a second run until we were delivering for three different companies which, in the world of the pamphlet, was a big no-no. But we were on good money for ten- and twelve-year-olds, earning about $20 a week for two deliveries. We got up to $40 a week when we took over the second round and one month, near Christmas, we made more than $900. That was the pinnacle of my pamphlet career. I finished the pamphlet round about twelve months before I went on the solo trip. The money from the job helped to finance my early adventures. Delivering pamphlets was one of the small steps in achieving my goals. As Paul Kelly says in his song, 'From little things, big things grow.'

∼

Things changed, as they invariably do, when I began secondary school. In 1994 I entered Year 7 at Wesley College, a large, well-known private school in Melbourne. It has a proud history as a boys' school and only recently threw open its doors to girls. I went to one of the newer campuses in the eastern suburb of Glen Waverley, an hour on the bus from home. I knew little about the school, except that it was big, students had more responsibility and freedom, and the uniform was purple, bright purple. I'm sure it was a tactic to prevent students from mucking about in public after school because everyone knew the distinctive Wesley uniform.

I think I suffered from the classic syndrome of big fish in a small pond who suddenly became a small fish in a big pond when I went to Wesley. I was house captain in Grade 6. At Wesley I was one of 180 Year 7 students, with a support base of one other kid from my old school. I was never part of the 'popular' group that was elected school captains and class monitors. I realised I was swimming in a different school of fish.

I was an average student at secondary school as my mind gradually turned to other, more interesting things. In my final year of primary school, Dad had returned from the Daintree to live in Brighton, a bayside Melbourne suburb. He told Beau and me about a sailing trip he and a friend had planned to make from Cow Bay to Cape York, the northernmost tip of mainland Australia. Not long after, the plan fell through, so Dad asked if Beau and I wanted to tackle it with him—would we like to sail more than 600 miles on a small catamaran, sleeping in the open and cooking our own food?

Would we what! Who wouldn't want to go camping with their Dad for two months in the tropics? I nearly burst with excitement at the thought. But, as I was only twelve at the time and Beau ten, Mum and Dad decided we had to wait a couple of years.

However, the damage was done. By the time I hit secondary school, my mind was away on the waves. All I could think about was adventure. I wanted to learn everything I could about sailing, which left me with little time to worry about essays and maths tests. I was passing, but I could have done a lot better, and I think the teachers knew that. The marks were average, but the reports were negative.

Mum initially had a few reservations about the Cape York trip. She thought it was a big responsibility for Dad, who hadn't had a lot to do with us up to that point, to take us away on his own for two months in what would be pretty tough conditions.

The First Steps

But Beau and I pushed to go, so she threw her full support behind us.

What most concerned friends and family was that Dad had little sailing experience. He didn't even have a boat! We went to the Port Melbourne Yacht Club every Saturday for a few weeks to learn to sail. I loved it. Moving under the power of wind was a new experience, and very different from the motor boats we'd mucked about with in the Daintree. A sailing boat is at the mercy of the wind and the waves, and has to work with them to move along. The motor boat, on the other hand, just thumps its way through the chop.

After completing the course Dad bought a secondhand 14-foot Caper Cat catamaran for $1000, with storage in each hull. It was the same size as the small catamarans for hire at any beach. It took Dad eight months of hard work to get it ready for our trip. I bided my time at school while dreaming of tropical beaches and lazing in the blazing northern sun. No wonder school was doing little to excite me!

To get some experience for the Cape York voyage, we decided to do a trip from Mornington (on Port Phillip Bay) to Barwon Heads, on the Bellarine Peninsula, about 25 miles to the southwest. On a map the journey seemed simple enough, but what a map can't show you is the treachery of the Port Phillip Heads. The Heads are internationally renowned as one of the most treacherous port entries in the world. The pressure of water gushing in and out of the narrow two-mile mouth at peak tides can create incredibly dangerous conditions. There are tales of 4- to 5-metre cliff faces of water suddenly appearing, or menacing whirlpools that could spin a 30-foot yacht out of control onto rough jagged rocks.

The sail from Mornington to the Heads proved to be far removed from my romantic daydreams of lazy days in the sun.

It was bloody hard work. And scary, because it was all so new. We made good progress but were getting drenched. I don't know if it was the sensation of a new experience, but I was pretty worried as we sailed across the Bay. We seemed to be going too fast, and the waves were crashing into us. And really, despite having taken a course, none of us were experienced sailors. Sailing on the sedate waters at the other end of the Bay was far removed from the conditions we experienced on the way to the Heads. We were all a bit green around the gills. But we eventually made it to the Heads, travelling the 20 nautical miles in five hours. We were absolutely drenched and close to exhaustion, but we made it. However, we knew that whatever we'd experienced so far paled in comparison to what lay ahead.

The gods were pretty kind to us, and we sailed through the Heads without incident. But the swell was amazing. There were few waves, but the swell rose and dropped, creating great moving slabs of water that we rode up and down. The sheer length of each swell was simply awesome, even if they weren't too steep. To be on such a small craft made it even more exciting.

We were soon in Bass Strait, which boasted a fearful reputation. The currents of the Roaring Forties that race around the bottom of the globe find themselves suddenly forced into this narrow channel between the mainland and Tasmania, creating incredibly fierce conditions for boats. Luckily, conditions in the Strait were also at their best and we made the short distance to Barwon Heads safely.

We camped at Barwon Heads for a few days, marooned by unfavourable weather. Our return trip was notable for a lack of wind, which forced us to travel the last stages in the dark. We had to guess which lights were Mornington, and missed our mark by many miles, which meant Dad had to hail a taxi to grab the car and trailer. (We found out later that the local

The First Steps

authorities were getting anxious about the car and trailer that had been sitting at the boat ramp for four days.) We'd passed our first serious sailing test, but there was enough to suggest that the Cape York trip would not be all fun.

∼

We departed in September 1995, a few weeks after my fourteenth birthday. Beau was twelve. I'd been waiting for this moment for nearly two years, which is a hell of a long time when you are fourteen. The trip was to take three months, which included a month to drive the 7000- kilometre return trip. You can imagine our excitement as Dad, Beau and I set out!

We hitched the Caper Cat behind Dad's car and left Sassafras for Cairns in the first week of the school holidays. We camped on the way up, stopping where we could, eating mangoes as we got into the tropical areas, washing and brushing our teeth in the rivers. We stopped at Brisbane to get some equipment, including a .22 rifle and some ammo—Dad wanted the gun to hunt wild pigs for food and also for protection in case one of the pigs wanted us for dinner. The only thing we actually shot was a coconut. I can't remember if we were hunting it, or if it was set to attack us.

After one and a half weeks on the road, we hit the warm humid rainforest of Cow Bay. It was a world away from the classroom. We sailed from Thorntons Beach the next morning, an overcast day, and hardly the ideal start for a tropical adventure. But pretty soon the clouds cleared to reveal one of those perfect tropical paradise days you see in brochures.

There was only a gentle breeze so we glided along the pristine beaches of Northern Australia, with its native trees and miles of untouched sand. Dad sunbaked naked; Beau and I mucked about with the fishing lines out the back of the boat. We caught two mackerel and I remember thinking, this is perfect. I look back at the moment and still think that.

We sailed for about six hours on that first day, and stopped for the night at a place called Cedar Bay. As we pulled in, a huge mackerel jumped out of the water and did a rainbow arc just off the bow of the boat. It was a fitting conclusion to a spectacular day. This is going to be a dream trip, I thought to myself.

On the beach, a man in the distance turned and walked away from us. This was one of the most remote places in Australia, so people usually go out of their way to greet other humans. But this guy just disappeared into the bush before we could get close to him. I always wondered who he was, and why he disappeared.

I later heard the story of a man called Michael Fomenko who would be about 60 years old. He had gone to live in the area in the 1960s, and was known to the locals as 'Tarzan'. Legend has it that he had been dragged off by authorities and 'lobotomised' many years ago. He was a simple bloke who ran everywhere and embarked on major adventures, once reputedly paddling by canoe to Papua New Guinea. He was a bit like the character Forrest Gump in some ways. I wonder if it was Tarzan we saw at Cedar Bay?

That night we cooked the fish we caught and ate it with rice. It was a pretty good meal, despite our poor cooking skills, and capped off a fantastic day. But nothing stays perfect forever. It started to rain before we had our tent up. That put a dampener, literally, on things. The next morning Beau hacked into a coconut, so we had coconut juice, cold rice and fish from the previous night's dinner. Cold burnt fish with gluggy rice was not a very appetising start to the day.

The food was the only real low point of the trip. Man, it

The First Steps

was awful. After a while Beau and I couldn't stand it. We moved from eating uncooked rolled oats mixed with water to swigging chilli sauce from the bottle just for the taste kick. By the end of the trip we were eating hot and spicy concentrated tom yum soup mix out of the jar.

From Cedar Bay we sailed on to the Hope Islands, then Cooktown, where we had the first scare of the trip. After leaving the shelter of the headland we were hit by a sea wind that caught us off-side. One hull abruptly lifted out of the water. Dad quickly uncleated the main sheet and the hull splashed back into place. It's quite common to lift the hull on a small catamaran, but it's not advisable to do it in an area infested with dangerous saltwater crocodiles. Who knows what was lurking nearby if we'd tipped over and been thrown into the water.

The cat had no cabin or shelter, forcing us to sit on the trampoline strung across the hulls, which left us exposed to the elements. There was simply nowhere to hide from the blazing sun. My nose got so badly burnt I still have to take extra precautions to ensure it gets no sun. I think that part of my face will need special attention for the rest of my life. Our hair went white blond, and our tans were on a par with aboriginal skin. We must have made a hell of a sight, this strange man on a slip of a boat with two small dark-skinned blond boys.

But it was the salt that really got to me. It invaded every part of our life. After getting drenched, and then dry, we would be covered in salt crystals, as though we'd been coated with a spray gun.

Most of our days of sailing were spent in silence as we tended to the boat and watched the coastline. There was not much to talk about, except when something exciting happened, like spying a 3.5-metre tiger shark gliding a few metres from us, or when Beau fell asleep and plunged over the side, waking to find himself dragged along in the water with his foot stuck in the rudder.

There was no timetable or planned route. The only requirement was that we finished by Christmas, when the wind turned from the southeasterly that was pushing us along, to a head-on northerly. We had no such worries, as we travelled an amazing 120 miles on our final three days on the water.

The long days of sailing placed a strain on all of us. There was also extra pressure on Dad because we were young. Before we left he laid down the rules of the trip.

'Everyone wipes their own arse,' he told us. We had to be responsible for the equipment, and work together to make things happen. It caused some friction among us, particularly between Beau and Dad, as Beau is a bit more headstrong than I am. A few times Dad pulled us into line, which I think we needed, but at the time we thought he was on our case. We were pretty young to be doing the trip, and were behaving like normal teenagers, leaving stuff behind or forgetting to do things. At home it may not have been a big deal, but on a trip like that careless behaviour could threaten our safety. I can actually see a bit of my father in myself now—I'm not very tolerant of people who do something wrong, or who don't respect property.

One of our stops was Lizard Island, an exclusive resort island 240 kilometres north of Cairns. It had fantastic beaches and coral reefs, making it ideal for scuba-diving, snorkelling and game fishing. As a consequence, it attracted some of the most amazing and luxurious yachts you'd ever find in the one place. In such surrounds we must have looked a sight—a bedraggled man and his two waterlogged kids, trudging up the beach. We were allowed to stay on the island on the condition we didn't

The First Steps

mingle with the guests. Nonetheless, we met a wealthy man who thought it was crazy of Dad to take his two young sons on such a foolhardy adventure. He insisted we take his Emergency Position Indicating Radio Beacon with us in case we struck trouble.

'I don't agree with what you're doing, but take this for the sake of the boys,' he told Dad. An EPIRB costs a couple of hundred dollars, so it was a pretty significant gesture. Luckily we had no need for it, and posted it back to him when we returned home. I hope he didn't need it while we had it.

I'd always been taken by the idea of adventure and there I was, on one of my own. I certainly did not want this trip to be my last, so I spent those long days on the catamaran dreaming about what to do next. I made up my mind on Lizard Island. The place was swarming with fantastic yachts —a marked contrast to our situation, floating about on two slivers of fibreglass eating burnt fish and getting our salt-encrusted bodies fried to a crisp.

Yachts were the way to travel, I reasoned as I inspected every yacht I could peer into. They had nice comfortable cabins, with all the fresh water you could want, were clean and were moored in a beautiful tropical location. That's when I started to think, hey, I'd love to get my own little boat, do it my way, go around the world doing all the stuff we were doing, but in a bit more comfort.

I could imagine sailing to exotic ports, meeting people, swimming in crystal-clear water, diving and catching fish. It would be fabulous. The irony was, when I did sail around the world, I never sailed to any of those places or met any people, except for a couple of fishermen. And as for luxury, people shudder when they see inside *Lionheart*, my home for eleven months.

After we left Lizard Island I told Dad I wanted to sail around the world. 'OK,' he said, as if I'd announced I was going to the shop to buy an ice-cream. Either he didn't hear me or he didn't take it seriously. Not that it worried me. My focus had shifted to organising my trip. My first priority was to get my hands on a copy of *Trade-A-Boat* magazine, which has advertisements for used motor and sailing boats of all shapes, sizes and budgets. If I was going to sail around the world, I figured I'd find the type of boat I needed in that magazine.

I couldn't get a copy of the magazine until we arrived at Thursday Island, where we completed the trip in December. In the meantime, I spent my days on board the cat drawing plans for my dream boat, working out where I'd put the supplies, making lists of food and equipment, and dreaming of a life sailing around the world.

We finally made it to Cape York two months after we left Cow Bay. It ended in a sort of comedy as we searched for the signpost to tell us we'd reached the tip of Australia. We finally found it so we could get the obligatory photo to prove we'd arrived. The next day we sailed the short distance to Thursday Island, where we stayed for a week while we waited for a ship heading back to Cairns that could take us and our boat home.

Almost immediately Dad began to plan another catamaran trip, this time to the Solomon Islands, about 1000 nautical miles east of Cow Bay. That trip is still in the planning.

~

I started the 1996 school year with my mind clear on one thing: I was going to sail around the world. How and when were yet to be worked out. My trip was virtually the only thing I could think of. No wonder my school marks were less than spectacular. On cold winter mornings, as I made my way down the

The First Steps

mountain on the school bus surrounded by screaming students, I'd put in my earphones, listen to music and look at the sun and think, 'It's a beautiful day today. I can't wait until I'm out there sailing around the world.' The notion of doing it non-stop hadn't been considered at that stage, until something happened on the other side of Australia.

In February 1996 seventeen-year-old David Dicks left Fremantle on his 34-foot yacht, *Seaflight*, to attempt to become the youngest person to sail solo, non-stop and unassisted around the world. The first time I heard of David Dicks was when Beau rushed into my room in late 1995 and yelled, 'There's a kid on the Today show who's going to become the youngest person to sail around the world.'

The news came as a total shock to me. I'd already started planning my trip in my mind. Little did I know that another boy, three years older than me, was planning the same thing, but doing it solo and non-stop. As David was in Perth, we didn't get much news of his trip on the east coat. I remember him leaving, as there was a big fuss when he turned back after a few days because of equipment failure.

A family friend, Phil Carr, who had taken our family on my first sailing trip when I was eight, and actually helped Dad prepare *Lionheart* in the frantic days before I left, gave me some books about people who had sailed around the world. There was one by Tania Aebi, an eighteen-year-old from the United States who in November 1987 returned from a two-and-a-half-year solo journey around the world on a 26-foot yacht. Another book was *First Lady*, by Australia's most famous woman sailor, Kay Cottee, who became the first woman to sail solo, non-stop and unassisted around the world, completing her journey in 1988. I got that book from the school library and was staggered to find it was signed by Kay. Wow, here was someone who'd done what I wanted to do and had signed the book I was holding.

What struck me was that the only sailing accounts were from those who had done a solo voyage and set some sort of record. It made for pretty inspiring reading. So inspiring that, as I read each book, the idea that I should attempt a solo voyage began to form in my mind.

The idea of being on my own, in control of my destiny and master of my boat was awesome. By the time I finished reading those books, my decision was made: I wanted to become the youngest person to sail around the world. American Robin Lee Graham was the youngest to commence a round-the-world trip at sixteen years old, so I decided to leave when I was fifteen, which I would be in August that year. I would need nearly a year to prepare, which meant a departure date of early 1997.

I met John Hill around this time. Dad had taught me to sail, but I wanted to know more. Mum had heard through a friend at work about a bloke who lived about five minutes from our house who had done a fair bit of sailing, and would make a good teacher. He did. His enthusiasm could knock the wind out of anyone's sails, and he knew what he was talking about. He'd actually just passed his written exam for Yacht Master Offshore when I met him.

Each Thursday evening, while most of my friends were at sports training, or doing homework, I would go to John's house for a two-hour lesson on navigation, ocean seamanship and ocean survival. He was a hard bastard who would grill me the whole time I was there. We'd go out onto the balcony, where John would have a smoke and a drink while I tried to solve problems he threw at me. He was one of my strongest supporters when I decided to attempt to become the youngest to sail around the world.

The First Steps

'Aim for it. Who knows,' he said.

I had nothing but a dream and determination, but I needed more than that. Specifically, I needed money to buy a boat. About $80,000, I reckoned. To raise those funds I needed a sponsor. I designed a logo and letterhead and sent about 50 letters to companies such as Uncle Tobys, Buttercup, GIO and Coca-Cola asking for financial support. I ploughed $75 in stationery and postage into the project. Here's what I wrote:

17 March 1996
The Managing Director

Dear Sir,
My name is Jesse Martin and I am fifteen years of age. I am currently preparing to leave on a quest to become the youngest person to circumnavigate the world alone by yacht.

The purpose of this letter is to invite _____ to sponsor me, either financially or in other ways in this endeavour. I am imagining that prominent company logos will be displayed on my yacht for the sake of all media coverage, although I am sure we can discuss these sorts of details at a later date.

If sufficient sponsorship is obtained I intend to depart from North Queensland early in 1997, and I estimate that my sailing time will be 170 to 180 days at sea (averaging 6 knots as set by Tania Aebi on a similar-sized boat), with the entire voyage taking one year. I will complete the journey at the age of sixteen and a half.

I have chosen a suitable boat which is the New Offshore 30. It's [sic] structure is sound in terms of design and strength and could well carry me around the world to success. I am also currently preparing myself with the necessary training including the psychological and physical aspects.

I am prepared for the responsibility of representing Australia and I have faith in myself and I ask that your company does as well.

I am looking forward to hearing from you at your earliest possible convenience.

Yours faithfully
Jesse Martin

I waited each day by the letterbox for the offers to roll in. I got 24 replies. Not one expressed any interest, except one business that sent me a discount voucher. And to really make me feel bad, three of the replies began 'Dear Ms Martin'. That was a real kick in the guts. Not only did they think I couldn't do it, they thought I was a girl. My trips to the letterbox were conducted with diminishing enthusiasm, until it turned to downright dread.

The letters for sponsorship may have sounded like a pie in the sky scheme for a fifteen-year-old, but I was serious. If I was able to get some support, I'd have certainly done it.

To comfort me in my disappointment, I turned to *Trade-A-Boat*. I think I was addicted to that magazine. It was in its pages that I discovered my next avenue for adventure.

It was an advertisement in the 'Crew Wanted' section. After my rejection I just desperately wanted to sail the world by whatever means I could. The solo age record was not as important as the actual adventure. If I could go with someone else I definitely would. I figured that it would be better to sail on

The First Steps

someone else's boat to get the experience for later trips. My only sailing experience had been the Cape York trip on a 14-foot catamaran.

The advertisement called for crew on a boat sailing from Brisbane around the world. No experience was necessary, but each crew member needed to pay his or her own way. This was it, I thought, as I hurriedly wrote my letter.

The ad had been placed by Cameron Smith, a 30-year-old ex-naval seaman from Brisbane. He was at that stage working in Tasmania as a builder. He was heading back to Brisbane and stopped by Melbourne on the way. Mum, Dad and I met him as he got off the *Spirit of Tasmania* ferry, which travels between Devonport and Melbourne. Cameron was the typical knockabout Aussie lad, with a stocky build and dark weathered skin from working outdoors. To complete the picture, he drove an old Land Cruiser ute with two big dogs in the back. But behind the rough exterior hid a nice smile and a pleasant manner, which helped to convince parents that he could be trusted to take their son away. We spoke about the other crew members, the proposed route, what it would be like. Mum was naturally keen to check him out to make sure he was not a whacko. I was only fourteen and this bloke twice my age was seriously considering taking me away for more than six months to the most remote reaches of the world. Any parent would approach the issue with plenty of worry. But he seemed to pass Mum's test.

He hadn't bought the boat at that stage, but planned to leave in December 1996 with four navy mates. Mum said before she would give the OK I needed to organise my schooling, which I promptly did through the Victorian Distance Education Centre. As far as I was concerned, that was it. I was ready to go.

When Cameron finally got the boat Dad and I went to Sydney, where he bought it, and helped him sail it to Brisbane with two of the guys who were to go on the trip. It was a 40-foot steel yacht named *Uomi* ('you owe me'). The trip took two weeks and we had a terrible time, with a headwind the whole way, which meant we hardly did any sailing, motoring most of the way. I was seasick, which was not a very good sign. This was the first real yacht trip I'd been on and I was as crook as a dog. It definitely wasn't how I imagined it would be.

It was on the trip that I heard of David Dicks again on the radio, as he was only two weeks from home. I remember being astounded by the 20,000 people who turned out to greet him when he docked in Fremantle.

We finally got to Brisbane and I was bloody glad it was over. But I was still keen to do the trip.

I headed back to school, where the trip with Cameron became my only thought. I'd constantly ring him, pestering for the starting date of the trip.

'Yeah, not ready yet, almost ready, almost ready,' he'd say each time. This went on for months until he rang with the bad news. The others were having trouble financially, so the trip had been postponed.

I was disappointed as I'd pinned so much on that trip. Cameron was my ticket to complete a dream yet, despite his best intentions, he had whisked it away from me in those few words. After I hung up the phone Mum asked if I was disappointed. Yes, I was, but I decided at that point to take things into my own hands. It was obvious you couldn't rely on other people. I remember telling Mum, 'If I'm going to do this, I'll have to do it on my own.'

CHAPTER 2

From Belize on a Breeze

I was going to sail solo around the world. That much I knew. But realism was also creeping into my thinking. If I was going to do it by myself, I might have to wait a few years, perhaps get a job when I finished school to earn the money to buy a boat.

But the desire to do something was so strong, and school was driving me mad. I was in Year 10 and had become interested in documentaries, which I figured would be a good way to finance a round-the-world trip. But there's not much of a market for documentaries on kids going to school on a bus, so I began to plan my next adventure.

My mate Ben Richardson and I came up with a plan to kayak the Roper River, along the southern edge of Arnhem Land in the Northern Territory, after I read the book *I the Aboriginal* by Douglas Lockwood, which was based on the river. I'd been interested in Aboriginal culture since I was a kid after visiting Central Australia a few times.

We began to research everything about the trip, including where the crocodiles were and how to avoid being eaten by them. I also investigated cameras and filming techniques, of which I knew very little. My pamphlet job came in handy, as

I'd rip open the bundles and read all the electrical retail catalogues that featured camcorders. I also hung out at camera specialist shops, until I soon knew the best cameras for what I planned.

Everything was going along nicely until I read an article in *Australian Geographic* magazine about a sea kayak expedition in Papua New Guinea. I absolutely loved the shape of the kayaks in the pictures—they had such beautiful lines. (Only a boat lover would understand.) What made it more special was the setting.

One photograph in particular captured my imagination: it was a picture of a man in a kayak in beautiful crystal clear water, with native kids in their canoes surrounding it. I thought to myself, this had to be better than a muddy river in the Northern Territory.

I'd also heard from people we'd met on the Cape York trip of the remote parts of Papua New Guinea, north of the main island, surrounded by untouched white sand. Papua New Guinea won the day as destination of choice, as I desperately wanted to go to somewhere remote for this documentary. To be a true adventurer I had to go where few had been before. If I bumped into other boats or kayaks, that didn't count as an adventure.

Ben and I began planning the trip. My first step was to join the local sea-kayaking club, where I learnt to do the Eskimo roll, the 360-degree roll you see every kayaker do. Ben, however, was making little headway in terms of his preparations, so his involvement began to wane until he finally pulled out.

Beau was keen to do the trip, so he jumped aboard. I'd bought a yellow plastic kayak, but as I learnt more about kayaks I began to appreciate the fibreglass models, especially the ones designed by Larry Gray. Coincidentally, Larry had done a

From Belize on a Breeze

television documentary on a sea-kayaking trip in New Guinea that sold me on both the design and the destination. I found a secondhand kayak in Sydney on the Internet for $1050, which was relatively cheap for this design as they usually sold for $2400 new. A friend had a look at it for me and gave it the OK, so I bought the kayak sight unseen with money from my pamphlet round, and had it shipped down to Melbourne. Beau bought my other kayak off me, so we were set as far as vessels went.

It was time to seek some sponsorship. I prepared a seven-page proposal. Here's a bit of what I wrote:

> This expedition is focused on collecting footage on a digital video recorder to make an exciting documentary of the adventures and people that my expeditionary party encounter along the remote coastline of New Ireland, an island off New Guinea.
>
> The thing that will bring an exciting element to my video is the fresh approach of me, a fifteen year old with an inquiring mind and a zest for adventure. Because a sea kayak doesn't hold very much, our food will consist of what we catch, lentils, rice and other selected supplies. We will also be carrying basic camping and expeditionary gear.
>
> No motors or pollution: just two sensible boys battling the elements and reaping the rewards, leaving nothing but footprints and taking no more than recorded memories from along the picturesque coast line of New Ireland. I feel that this is the kind of place that people might want to see and an adventure they might want to experience on their television screens.

After outlining some technical details, I continued:

> As quite the adventurous type, I had planned to crew on a yacht, circumnavigating the world this year. I anticipated filming this trip to raise funds for my own personal endeavour, namely, to become the youngest person to sail solo around the world and if this is to happen and I break my records, I must leave at the start of next year.
>
> The opportunity to crew on this yacht was sadly postponed for twelve months and my father's proposed trip to the Solomon's on our Seawind 24 is still in the making. It is now that I am left to find other means by which to raise the necessary funds for my solo quest. I hope to do this as a result of my kayak expedition to New Ireland.
>
> Relying on other people has met with disappointment. Therefore I need to take control of my own adventures to avoid any more further setbacks.

I tried the same companies I sent my letter to the previous year. The first sponsor to come on board was *Australian Geographic* magazine, which gave us $3000. In return I would do a story and Beau would take photos for the magazine. I was rapt. The clothing company Snowgum gave me some equipment, and Air New Guinea gave me a good deal on the airfares. QBE provided the travel insurance and Quicksilver gave us some clothes.

But all this help didn't mean the trip wasn't a financial burden. I bought a broadcast-quality camera that a friend, Paul McLellan, got for me in Singapore duty free for $5000. I also needed a special housing to film underwater. Funnily enough, the bloke who custom-made the housing lived in Kaniva, a small town in the wheat-belt on the South Australia–Victoria border. It is literally hundreds of kilometres from the beach, making it one of the strangest places to find this sort of equipment. We drove five hours to look at the casing, stayed an hour and drove all the way back. We did that trip twice. I ended up renting the casing off him, as it was too expensive to buy. I also bought sound equipment, tripods, cases and all the bits and pieces that go with filming. The trip cost me $9000, of which I borrowed $4000 from Mum.

Besides chasing sponsors, the preparation involved things such as planning the route, researching the best time of the year to do the trip and working out how to ship the kayaks to Papua New Guinea. I worked on planning the trip most of the year, which made a fair dint in my studies. I actually did much of my research at school, particularly during subjects such as information technology, when I could look on the Internet while I was meant to be learning some computer jargon. I had to—I knew nothing about Papua New Guinea when I began planning.

Beau and I were meant to leave mid-year, but we were not ready, so we didn't leave until the final term holidays. The trip took five weeks, which meant we ate into three weeks of the final school term, which was a pity.

Unfortunately school had become a bit of a distraction by then. I did enough to get through, but I could tell the school was frustrated with me as I could have done much better. I actually never raised the fact that I was planning a trip, but I believe the teachers knew what I was doing. I never asked for any concessions, and didn't receive any. My friends, of course, knew about it, but I don't think they expected me to go through with it. Many kids say things to their mates without following through. Plus,

I don't think they could understand why I wanted to go on a trip that involved a lot of work and a fair dose of hardship and suffering.

I may have been planning the kayak trip, but I never stopped preparing for my solo trip. In July I made my annual pilgrimage to the Melbourne Boat Show, where I'd scurry from stand to stand, researching, inspecting equipment, and familiarising myself with every new boat and gadget I could lay my hands on. I was one of those annoying kids who collected a brochure from every stand but, unlike most people, I actually read them when I got home.

I was on the prowl at the show when an announcement came through the loudspeaker that David Dicks would be appearing in five minutes. The idea blew me away. There, in the same building as me, was the guy who had done what I wanted to do. I had to hear what he had to say and to meet him. I left Dad talking to a man about parachute anchors and began scouring the site for the place he'd be talking. I went up to one stand where there was a group gathered around, but it turned out to be an old guy talking about fishing. Time was getting on, and I was getting very anxious. I ran back to Dad, who was still engrossed in his discussion on anchors. I then realised the stage where David Dicks had appeared was just a few metres from the spot I'd been standing when I heard the announcement. It was obviously over. I was devastated at missing my chance. But as I was about to walk away, I noticed one person still at the stand, who appeared to be packing up. It was David Dicks.

Dad and I approached him and introduced ourselves. He turned out to be a normal person, like you and me. He struck me as a real sailor, which is hard to explain. He was not into the record, or the specifics of the trip, but just wanted to talk about sailing and the gear he used.

From Belize on a Breeze

He happily flicked on his documentary and then chatted to us and signed his book for me. Dad did most of the talking, because I suppose I was a bit in awe of my hero. I went home on cloud nine, and more determined than ever to do the trip. I had met the youngest person to sail around the world solo, and he was just another human being. I knew it was possible.

~

The Papua New Guinea trip caused a bit of a ruckus in our family. Mum insisted on coming with us, as PNG had a reputation as a violent place, particularly the capital Port Moresby. Dad was adamant Beau and I must do it on our own if we were to keep it pure in an adventure sense.

He never spoke about it, but I know Dad felt his parents didn't support him in a lot of things he wanted to do, which was why he had such a strong reaction to Mum's plans. I didn't want Mum to come—I felt it wouldn't be an adventure if an adult came along. But I can see her point now. I'd just turned sixteen and Beau was fourteen. It was a big leap of faith for any parent to let their children travel unescorted for five weeks to a place such as Papua New Guinea. She wasn't worried about the kayaking trip, just the travelling through the main cities to get to our starting point.

In the end Mum and her partner Andrew came along. We flew to Port Moresby, which was a bit of a culture shock with its ramshackle airport and strange smells. After watching some locals eye our backpacks, we were pretty glad to get our connecting flight to Rabaul, in the province of New Britain, northeast of the Papua New Guinea mainland. This was the starting point of our trip.

We hoped to greet our kayaks, which had been shipped ahead, and leave as soon as we could, preferably the following day. But, as is often the case in remote places, the kayaks didn't arrive for another one-and-a-half weeks. At least it gave us time to explore Rabaul. It is an amazing place, established in 1910 as the capital of German New Guinea, but is probably best known for its role in World War II. The city was invaded by the Japanese and used as a major submarine base for their Pacific fighting. Relics of the war can still be found, and the locals are known to use gunpowder from old bullets to make fish bombs.

Our plan was to begin the journey at Rabaul, and paddle to New Ireland, across St George's Channel. We'd island-hop across to Duke of York Island and the surrounding islands that dotted the channel. From there, we'd follow the New Ireland coast north, pulling into villages until we got to our destination, the tropical town of Kavieng on the northern end of New Ireland. I'd chosen New Ireland, which was part of the Bismarck Archipelago, as it was one of the most remote places you could actually travel to in the Papua New Guinea circle of islands.

When we arrived at Rabaul we were warned not to cross St George's Channel as a typhoon to the north, near Japan, was sucking wind through the straight. That, combined with the strange weather patterns from El Niño, which had a pronounced effect on tropical areas, had caused unusually rough weather with a steep chop in the channel. We were told that many boats had disappeared without trace in the last few months, presumably caught in a current and dragged out into the Pacific Ocean. Imagine what the locals thought when two young teenagers arrived to attempt the crossing in kayaks! Luckily, the day the kayaks arrived, the 24-hour forecast was for exceptionally light to no winds. We had to leave the next morning when the conditions were calmest.

Before we left Rabaul, we were lucky to have our first

encounter with Papua New Guinean islander culture. I'd read about the traditional ceremonies in many of these areas, so when the chance came to witness a *'Sing Sing'*, the Papua New Guinean term for celebration, we jumped at it.

That evening we took the bush track to the centre of a small village, following the drone of beating drums and chanting, which grew louder the closer we got. Suddenly, a piglet squealed as it darted out in front of us. It seemed so strange, yet entirely appropriate in the circumstances. I felt like an intruder walking with my camera into a situation that to me was completely new, but had been a way of life for these people for thousands of years. A part of me felt a bit like a true adventure documentary maker, recording a final frontier.

We had joined an initiation ceremony that had actually begun two weeks earlier. A group of young boys had been sent into the bush to learn the secret of black magic while under the power of the betel nut. This nut, when eaten, acts as a powerful drug, resulting in hallucinations and severe head spins. While they were in the bush they would get high on the nut, then learn the dance they were to perform at the ceremony we saw that night. The sound of the chanting and the monotonous beating drums sent them into a trance, enabling them to dance for tremendous amounts of time under heavy costumes. They emerged from a hut and began to dance in front of the villagers who beat drums and clapped to a haunting beat.

None of the women or children were allowed to know anything about what occurred leading up to the night. They remained in the background and watched under the protection of a stroke of lime on their faces. During the course of the evening a lady came up to me and, with her finger, stroked my cheek with the white substance.

While the dance was underway, the food, a major part of

these traditional ceremonies, was distributed according to strict tribal rules. The bananas, wrapped up over the last week to ripen, and the *kau kau*, or sweet potato, were divided into family groups. The uncooked pigs were cut up, with the best portions given to selected families. For a family to receive a pig's head was an immense honour, and they had to supply a pig for the next ceremony. The villagers took the rites very seriously, and those who dissented risked punishment by death if the rules weren't followed.

What really struck me about Papua New Guinea was how generous the people were. The country was suffering a horrendous drought, yet we were always given food, no matter how hard it must have been for the people to provide for strangers. At the ceremony we were treated equally, receiving some bananas and a portion of pork. (Luckily it wasn't the head.) That night, as I watched the ritual around me, I experienced something I'd never felt before, a very strong spiritual feeling that spooked me. I mean, this stuff was real!

Our trip finally got underway the next morning under the cover of darkness. The first stage, from Rabaul to Duke of York Island, was to be the most dangerous of the trip, where we'd encounter the worst of the currents that had swept the local boats out to sea. We planned to get to the island that day, spend a few days paddling its length before pushing on to New Ireland. But a few miles before the island, we struck trouble. The current became so strong that, despite hard paddling, we made little progress. Luckily we were in radio contact with Mum and Andrew, who had made friends with a local businessman who'd taken them marlin fishing on his boat that day. They were able to tow us to Duke of York Island.

We spent the night in a grass-thatched hut, cooking damper and sharing it with the locals. Two boys of similar age to Beau and me were assigned to escort us for the night. We offered them some of our damper, but they politely refused, preferring to stick with their fish. Early the next morning, the boys cut down some coconuts for us for breakfast.

From Duke of York Island, we headed to the north of the island. As we paddled along, a group of people waved to us, signalling for us to come in, which we did. The whole village wanted to shake our hands. The villagers chattered excitedly as they gave us *kulau*, a baby coconut not yet fully developed with sweet milk inside. It took us ages to finally get away.

As we passed more villages, villagers would run from their huts and beckon us to visit. But Beau and I had learnt our lesson—we'd wave back and keep paddling. If we stopped at every village, we'd never get anywhere.

I was keen to find some sunken World War II tanks off Duke of York Island that I'd read about. But when I saw the large shadows underwater, I wasn't too keen to take a look, as it was hard to tell if they were tanks or sharks. Luckily, they turned out to be the tanks.

I was terrified of sharks, and the feeling that they could be lurking nearby really played on my mind during the entire trip. I jumped into the water to get some footage of the tanks, but didn't stay underwater too long.

Our last night on Duke of York Island was at Waterhouse Bay at the northern tip of the island. We were lucky enough to be invited by the richest man in the village to have dinner with him. For him, it was a major honour to have visitors dine in his hut. After dinner, half the village appeared to gather outside our tent, chattering excitedly until a guitar appeared and the singing started. It was a beautiful warm night, like it usually was in Papua New Guinea, with the stars in full view. More and more lanterns appeared on the water as men took out their canoes to go fishing. We headed off to bed with the knowledge we must be up early in the morning before the seas built up in St George's Channel. It was an amazingly peaceful atmosphere. Moments like that renewed my commitment to get out to see the world.

Life moved at a different pace in Papua New Guinea. The villagers seemed to just hang around and do nothing. Their only responsibility was a few hours of gardening each day to keep the weeds under control and to maintain the fences to stop the pigs from destroying the gardens. These meagre responsibilities were put on hold if the urge to go fishing or have a snooze became too strong. If something like a new bush knife or pair of shorts were needed, they would simply collect some coconuts, cut out the copra to dry, then sell it to the businessman of the village or make a trade for the item they needed. They wanted only the bare essentials, so there was none of the greed or material desire we were used to. The western world could learn from that kind of mentality.

The administration of law and order was another thing that separated us, although I don't advocate that we follow their example. The only law in these areas was local law. If someone was caught stealing, he or she was killed, no questions asked. And this was in communities involved with the Uniting Church. It looked to me as though the locals were just attending church to keep the missionaries happy, so they wouldn't lose the support the Church gave them in terms of food. I often asked people how they could adhere to two apparently conflicting beliefs— one of compassion as taught to them by the Church, the other of brutal street justice and black magic. Not once did I get an answer. I don't think they really knew themselves.

A typical day started as the sun came up. The tent got too hot if we stayed in it any longer, so we had no choice but to start early. We nearly always had food left over that had been given to us. On the odd occasion we didn't, we cooked a bit of damper or noodles for breakfast. The coastal waters were rougher than we imagined, which got us into the habit of leaving early while it was calm and retiring when the sun was at full strength, at about 2 p.m., and the waves threatened to tip us over. We had no plan of where we would end up each night, merely selecting a friendly looking village as our stopover.

After saying goodbye to our new friends, we'd head for the furthest tip we could see and be on our way for the day. We usually paddled for eight hours daily, all the while filming everything around us, while Beau took some spectacular photographs. We'd see the most amazing wildlife. An entire school of flying fish would emerge from the water at once and hit the side of our kayaks with a loud thud. Huge manta rays would jump into the air in the distance, doing a double back flip, apparently just for the fun of it or perhaps to get a look at us. Because our paddles made hardly any noise, we could sneak up on unsuspecting turtles as they lay semi-submerged, sunning themselves. But when they saw us, they were off in a flash, moving a hell of a lot quicker than you'd expect a turtle to move.

But the wildlife was not always a pleasant surprise. One day we finally saw what I'd been dreading the entire trip. About ten metres from my kayak lurked the outline of the biggest shark I'd ever seen. Its torso would have been the size of a 200-litre drum. And it was heading our way. I yelled—as quietly as you can yell—to Beau to be quiet, stop paddling and get the paddles clear of the water. We tried desperately not to attract attention to ourselves, which was difficult, given that we were the only people crazy enough to be in this wild and remote part of the world, and we had the brightest fluoro blue and yellow craft and equipment you could imagine. As I was pointing out the lurking shadow to Beau it suddenly disappeared. This was worse, as I was paranoid it would surface next to us and take a huge bite out of our kayaks. We decided to make a dash for shore, about a mile away, which I'm sure we covered in record time. After a rest and enough time to build up the courage to head out again, we hesitantly continued on our way. But our eyes were on red alert.

We stopped at the village of Kontu for a few days' rest. There, we experienced something I'll never forget. I'd heard about shark callers and the traditional way they caught sharks, so I was keen to film them in action. The shark callers, mostly old men who still practised the art, would take their canoes out to sea and clap together two coconut halves underwater to mimic the sound of struggling fish to attract the sharks. When a shark approached, they would place a fish as bait on the end of a pole. When the shark lunged at the fish, they would slip a noose, attached to the pole, around the shark until it was trapped. The shark was then clubbed to death.

The first time we went out they had no luck, which was blamed on our presence. A second attempt saw a nearby boat nearly noose a shark, only to have it bump its nose on the boat and back away from the pole. The incredible thing about the shark callers was the superstitions they had. For instance, they were not allowed to sleep with their wives or nurse a baby for three days before going on a hunt. And to step on pig or bat droppings in the days leading up to the attempt would ruin any chance of success.

But it was the next day that we were in for a major shock. One of the villagers, who spoke English, revealed that a pod of dolphins had been trapped the previous day and the entire village was going out to spear them for food. The Papua New Guineans were in the middle of a horrific drought and needed the dolphin meat to eat. We were told that the previous day a group of men had gone out in canoes, tapping two rocks together underwater to attract the dolphins. They lured the dolphins into a shallow reef, which, at low tide, formed a natural pen. Once there, the entrance was blocked with dried coconut branches weighted down at the stalk. The fronds floated upwards to create a fence, which the dolphins, relying on sonar to guide them, could not penetrate.

The next day arrived, and the men waited in their canoes, spears at the ready for the dolphins to surface for air. The children would swim about, signalling to the men where the dolphins were about to come up. Each dolphin would be speared a number of times, until they slowed considerably. They would then be clubbed to death and dragged ashore. This process would continue until all the dolphins were killed. This may shock many people, and be looked upon as brutal and cruel, but we do basically the same thing in the western world, don't we? We kill cows to eat when we can easily live on grains and vegetables. Just because we get our meat prepared in a polystyrene container with plastic wrapped around it and don't see the slaughtering doesn't make us any different from them.

The people of Papua New Guinea were extremely spiritual. They believed that all things, good and bad, were under the control of the spirits. Malagan magic was the central belief. Everyone, from the village elders down, was terrified of the spirits, which the villagers believed dictated every moment of their lives. I met a young man, Michael, who became a good friend. He told me his mother had been killed by a man in their village about nine months before. Apparently this grumpy old medicine man, whom nobody liked, had a grudge against the family, so he stole a piece of Michael's mother's clothing, took it into the bush and performed black magic on it. He returned it before she knew it had been removed. When she next wore it, she fell violently ill for no apparent reason. She was rushed to Kavieng Hospital more than 100 kilometres away. Three days later she died of breast cancer. I asked Michael if everyone in his village knew what the old man had done. He said it was common knowledge, but everyone was too afraid to do anything because the old man would also kill them.

~

Beau's arm began to play up at this stage of the trip. A few years before he'd had a pretty bad skateboard accident, smashing a number of bones, which required a major operation. A number of plates and screws were inserted in his arm. Then, while he was in hospital, he contracted staph in the wound. He has had three operations for the breaks and the wound still weeps from the staph.

It began to play up about five days before the end of the trip. At the same time we had a chance meeting with Mum and Andrew. It was decided that Beau would hitch a ride on a passing truck with his kayak to Kavieng, while I prepared myself to finish the trip solo. But Mum really didn't want me to do it on my own. If only we hadn't bumped into them! Luckily, Michael agreed to accompany me on the final leg.

We travelled up the coast to a small village called Kaut. The next day we had planned to paddle through the Albatross Channel and arrive at our final destination, Kavieng, but the village chief strongly urged us, with Michael translating, not to paddle through the channel. The previous week some rascals had held up a boat and killed all on board to steal their cargo. They hid in the labyrinth of small creeks and mud flats where the police were unable to catch them.

The chief insisted we use his fibreglass dinghy, which had

an outboard motor, to reduce the risk of being held up. Michael convinced me this would be wise. He was very worried. The next day, after I gave the chief some fishing hooks and sinkers for the use of his boat, we lifted the kayaks aboard and jumped in as the outboard revved into life. We arrived in Kavieng after paddling 150 miles over two-and-a-half weeks. It had not been easy, and there were times when I was terrified, but I had found the adventure I had been looking for.

Mum, Andrew, Beau and I arrived home in late October. One of the first things I did was to organise a professionally produced five-minute teaser tape of the trip, which I wanted to tout to television stations. Unfortunately, things like that took longer than I expected, and other events soon took priority. We did the article for *Australian Geographic*, while the footage stayed in a shoe box under my bed.

~

I returned to school for the last couple of months of Year 10, and renewed acquaintances with my favourite publication, *Trade-A-Boat*. I was still searching for a boat I could crew on around the world, because I really needed that experience before attempting a solo trip.

I came across an Australian boat that had spent the last three years meandering around the globe. In a few weeks it would be in Florida, in the United States, where it wanted fresh crew to sail the final six-month stretch to Australia. I was keen. Very keen. But I had to weigh that up against the disappointment of the aborted attempt twelve months before. I was only sixteen years old and had done no real distance sailing. It was a long shot that the skipper would want to take such a young and inexperienced hand on board.

Crunch time had also arrived for my schooling. I was due to

start Year 11, the first year of the crucial Victorian Certificate of Education, in a few months. This was the first of the two-year Leaving Certificate, which I needed if I wanted to go on to university.

From the moment you can write two words, you seem to have drilled into you the importance of doing well in your VCE. If I was to embark on my final years of school, I had to make a decision. Was I going to continue chasing my dream of sailing around the world, or knuckle down and finish school? I could not afford to scrape through like I had been doing—the teachers would not be so tolerant in those final years.

But no matter how hard I tried to reason and listen to the arguments in my mind, I ached to go on another adventure. It was an opportunity too good to miss, so I applied for the position. It was a few weeks until I received a reply. I remember the nervousness as I opened the letter. I also remember the utter disappointment when I read the rejection. They'd filled the spot and no longer required crew. I was devastated. This had been my best chance, I told myself, and now it had gone.

The letter sealed my fate. I entered Year 11 in February 1998, determined to do the best I could. My life of adventure would have to be played out in my mind on those cold winter's mornings as I looked out the bus window on the way to school. I'd get out there one day, I reasoned, but it may not be for some time yet.

Then something happened, as it always does.

I got home from school about two weeks into the school year to receive a phone call. My heart skipped a beat when I heard the peculiar sounds that come with an international phone call. I swear I nearly fainted when the strange voice introduced himself. It was Dave Smith, the skipper of the boat in Florida. The crewman had failed to arrive. Did I want to join them? They would be in Belize, a small country bordering Mexico and Guatemala on the Caribbean Sea's Gulf of Honduras, in a fortnight. If I was going to join them, I had to meet them there, he said.

I was beside myself with excitement. I got off the phone after mumbling something about getting back to him. Then the reality of the situation hit. I'd already started school, and if I was to embark on the trip it would probably put me too far behind to catch up to make a proper go of VCE. Mum knew I desperately wanted to go, and would support whatever I decided. She had gone to the trouble of talking to Dave's mother to get an idea of what sort of person he was, so she was comfortable with the trip.

Dad, on the other hand, was against the trip. That surprised me as I was sure this would be something that he'd tell me to go for. He supported our Papua New Guinea trip, but I'm still not sure why he was against this trip. I suspect he was a bit concerned at granting so much freedom to a sixteen-year-old by sending me to a dangerous part of the world with strangers. I can understand how that would worry a parent. And I can see how anyone could get into trouble while they were in places like that. Dad's opinion was important to me, so it made my decision to go all the more difficult. I just couldn't pass up this chance.

I'd never tackled anything with such gusto as I did in preparing for that trip. I only had a bit over a week to organise my schooling and equipment for half a year at sea. The first step was to get in touch with the Victorian Distance Education Centre and organise study material. I then had to scrape together some money, as I had none left after Papua New Guinea. On top of the cost of getting to Belize, it was going to cost $150 a week to crew on the yacht to cover food and overheads. Once again, I had to borrow the money from Mum.

Then there was the organising of flights and visas. It was an incredibly hectic time but, with the help of good old Mum, I made it. Two weeks later, I said goodbye to my family and boarded a plane for Los Angeles.

But not before the obligatory dramas. When I presented my one-way ticket to Belize at Melbourne Airport, the immigration officer immediately summoned his superior for a mumbled conference, complete with concerned glances at me, then back at my ticket. They finally told me why they were worried. They believed I risked being turned back when I got to Belize as I did not have a return ticket. Even if I explained I was going to sail out of the country on a boat, the Belize officials may still not believe my story and turn me back. I had no return ticket to jump on a flight out of there, so I'd be stuck. They were also very concerned about sending a sixteen-year-old into a volatile situation like that, especially in a Central American country. As you can imagine, Mum was nearly having kittens by this stage. I ended up signing something that stated my intentions before I could go. I'd had too many disappointments to let some officials at an airport stop me. Mum even got a call from the Department of Foreign Affairs in Canberra to say they were concerned at what could happen, and would be monitoring my situation.

I got to Los Angeles, then on to Houston the next day. From there it was direct to Belize. I had about 40 hours of flights and stopovers to mull over my possible rejection at the end, yet, when I got there, they didn't even bloody look at me. I must have looked like a trembling leaf as I got off the flight, but not one Belize official batted so much as an eyelid at my one-way ticket. I quickly grabbed my bags before they changed their minds, and headed for the street. I was sitting in the gutter, eating some of the bubblegum I'd stocked up on in the United States, when this

From Belize on a Breeze

guy approached me. He was in his mid-thirties, his long hair spattered with blond tips, and wore the tan of a man who'd spent some serious hours outdoors. He carried a woven basket with some fruit and vegetables in it and was walking in bare feet through the filthy streets of a third-world country. It was Dave, my skipper for the next six months.

He wasn't much to look at, but I was bloody glad to see him. We introduced ourselves, then hailed a dodgy-looking taxi. He stood in this crazy street yelling and gesturing as he negotiated the price. For a split second I felt as though I was in a movie. Only two weeks before I'd been committed to two years of heavy study. My mates would have been doing homework at that very moment, and here I was, about to head to sea with this wild-looking bloke in a country I'd never heard of until two weeks before. If I was looking for adventure, I reckon I'd found it.

We arrived at the boat without being ripped off, thanks to Dave's negotiation skills. I was just so happy to have made it to the dock, to finally see the boat that was going to take me back to Australia. All the rejections and disappointment were well and truly behind me. The yacht was an Adams 40, a 40-foot fibreglass yacht built by Dave in 1994. He had sailed her pretty much ever since, around the world through Indonesia, Madagascar, South Africa, Brazil, America and the Caribbean. She was called *Imajica*, the title of a book Dave had once read. He thought it was a nice name, even though he didn't like the book.

I also met the crew: two Australian women, Moira and Anika, who were both in their mid-forties. They'd joined Dave in Miami after answering the same advertisement.

Meanwhile Mum, who was beside herself with worry, had rung the Belize airport to see if the Houston plane had arrived. In broken English they told her the plane had landed, that everyone had got off, and nobody was at the airport. But that was of no comfort to Mum. She was convinced they'd stolen my stuff and thrown me in the slammer. I obviously didn't share Mum's concern. It wasn't until after dinner, more than five hours after I arrived, that I realised I needed to ring her. At least I called her the day I arrived!

As the crew had been waiting for more than a week for me to get there, we left early the next morning. We left so suddenly we didn't go through customs or immigration. Belize immigration probably still thinks I'm in their country somewhere! But, then again, they probably never noted I'd arrived in the first place. It was the same in Panama. They certainly march to the beat of a different drum in Central America.

The first few days at sea we mainly stuck around the coral atolls about 50 miles offshore. It was the first time I'd really been ocean voyage sailing. The trip from Sydney to Brisbane had been mainly under motor, so there was a different feel and motion to the boat. I wasn't seasick, but there were definitely periods of queasiness as we headed into the waves. Some of the crew were physically sick, which usually makes you feel worse. Dave could just about sail the boat on his own, which meant our role was to help out when we could. I think I learnt more than the girls, and was quickly able to do more once we got underway. By the end of the trip I was the one he called on if he needed something done, which was reassuring. I had hoped to be doing all this myself one day.

It took a week at sea before we got to Panama and the famous canal that separated North and South America and linked the Caribbean Sea with the Pacific Ocean. The canal is not only an engineering marvel, but it makes journeys such as the one we were making possible. Without it, around the world cruises would have to take on Cape Horn, which would not be a very pleasant experience. The global pleasure cruise would be virtually non-existent without the canal. Ironically, nearly a year to the day after we passed through the Panama Canal, I was battling those very conditions at Cape Horn that we were avoiding.

Sailing through the Panama Canal was quite complex. We had to spend two weeks in Colon, an ugly town at the eastern end of the canal, waiting our turn and doing the necessary paperwork. As it was a busy passage, it was necessary to book a spot. At that time, the Americans were handing control of the canal to Panama, which meant the cost of the passage was expected to rise, so a lot of cruisers were trying to get through before the price hike.

To sail through the canal, you need a skipper, four lines people and a pilot, who works for the local authority. The crossing takes two days. The first day you pass through the first lock, which lifts boats up to enable them to enter the large freshwater Gatun Lake. After waiting overnight at the halfway point, a new pilot is dropped aboard, and you sail through the second lock, dropping back down to sea level to enter the Pacific. The smaller yachts have to help each other, with each crewing the others' yachts through the canal to Balboa, the port at the western end, then catching the bus for the 60-kilometre journey back to Colon to bring another yacht through. We needed one extra person, as well as the pilot, to get through. We ended up doing a few trips with other boats as we waited our turn.

It was in Balboa that I got mugged. Dave and I decided to catch the bus to Balboa to pick up mail, and watch a movie in the air-conditioned bus. We were walking through a rough part of town when three guys in their twenties came up behind us and grabbed my arm. They tried to grab my watch, which had a rubber band and just stretched, so I yelled, 'Have it. Have it.' They took the watch but still held on to me, which was a cause for concern, as I had my passport and money in my pocket. If I'd lost that, I would have been in real trouble.

Luckily Dave did some quick thinking. Only a few minutes before he'd bought a new pair of sneakers, which he had on. He was carrying his old shoes in a bag, which he intended to give to the next street hobo he came across. Instead, he yelled, 'Here, catch,' and threw it to the guy holding me. The thief naturally went for it, thinking it was something valuable. We just ran. It's a pity we never saw their faces when they opened the bag and got a whiff of Dave's smelly old runners. I was a bit shaken, but glad I had a good story to tell my mates when I got home.

We finally got permission to enter the canal. It was not as romantic as one might expect. The canal is a pretty bleak place, with plenty of concrete around. We were in the first lock when I noticed smoke coming from the motor. I yelled out to Dave, who seemed to dismiss my concern, until I pointed out the billowing smoke. One of the springs was damaged, causing trouble with the pistons. After doing some temporary repairs, we stopped at the halfway point for a few days while Dave caught a bus and a train to get a part for the motor. A two-day trip was stretching to three weeks, but at least we were in the canal. We finally got the motor repaired and waited for the next morning, when a new pilot would arrive. There was nothing to do but swim and wash and socialise with the other yachts nearby. One bloke jumped into the water and swam to our boat. He was sitting in our cockpit with a beer in his hand when Anika spotted a crocodile about 50 metres from the boat. We stopped swimming from that moment. To think some people have actually swum the length of the canal.

From Belize on a Breeze

We eventually got through the canal, and not a moment too soon. We'd spent nearly three weeks in Panama, when we really wanted to be sailing. We spent a night in Balboa, then got out of there as quickly as we could. Our next stop was the Galapagos Islands, famous for their volcanoes and wildlife.

~

We were at sea for eight days when we first spotted the Galapagos Islands, really prehistoric-looking islands. You half-expect a pterodactyl to fly out from behind one of the mountains. The islands were formed from volcanoes millions of years ago and have some really weird animals. This was because of the cold Humboldt current coming up the coast of South America from Antarctica to the Galapagos. So the island, located right on the equator and with an average temperature of more than 30 degrees, had animals usually found in cold climates such as seals and penguins.

The prehistoric bent is lent weight by the large marine iguanas. They look like huge lizards but actually live in the sea, diving and eating sea grass off the ocean floor. Galapagos was also famous as the place where Charles Darwin thought of his theory of evolution. He noticed finches on the islands differed slightly to suit the different conditions on each island. He reasoned that maybe the finches came from the same ancestor but evolved through natural selection to their current state. I actually found a dead Darwin finch on the ground and put it in my pocket to take home. But I forgot about it and proceeded to try on some wetsuits to go diving. When I got back to the boat I pulled out the finch but it didn't look very happy so I ditched it.

The wildlife was the most incredible part of the Galapagos. A couple of times we saw a seal wandering around on the street in the middle of town. From the reaction of the locals, seals in the main street was a pretty common sight.

The Galapagos was a beautiful place, but there was one bad memory from our visit, which made me wonder if I'd die.

Moira and I wanted to go scuba-diving, so we went on a tour to Gordon Rocks in hope of seeing some hammerhead sharks which, again, were only meant to be found in colder climates. The guides took us to a place that was quite ferocious, with white water spewing up around the rocks. It was a fairly angry sea to be diving in, but we jumped in anyway. I soon discovered my mask did not fit properly as water poured onto my face. Usually this is no great bother, as blowing through your nose would most often clear it. But this mask just would not seal.

I started to lag behind the group. As much as I looked, I couldn't find any hair breaking the mask's seal. I suppose I was not paying attention, and feeling a little tired from trying to clear the mask, but suddenly I found myself being swept along by an incredibly strong current. I was about 10 metres underwater, where the currents pulsate up to 6 knots an hour. For a current, that was incredibly quick. I found myself in the middle of a channel where the current swept through. I was only a few metres away from the shelter of a large coral clump, but the force of the current meant I was not going to make it. I was absolutely buggered, and the water was pouring into my mask, so I could not see the others in the group. I tried to swim, but was getting nowhere, and the harder I tried, the more I could feel myself being swept away from the group. I believed I was on the verge of being swept out into the open ocean, when I saw a rock below me. I dived and grabbed it, hanging on like Superman flying through the air. But now that I was stationary, the force of the current against me was much stronger, creating a bigger problem. On the mouthpiece was a button that

From Belize on a Breeze

allowed air to escape. The pressure of the current was enough to push that button in, purging my air.

I was in a bind. If I tilted my head forward and looked down, my mask would fill with water, and more than likely be ripped off. If I held my head up, I'd lose more air. I started to panic. My goggles were full of water, I had no idea where the others were, and I was losing air rapidly. As the air was being purged, my mouthpiece filled with water, until I was breathing a deadly mix of air and water. I thought I was down to the final mouthfuls of air in the tank. I made the signal for more air, not knowing if anyone was around me, imagining the worst.

I was never so glad to feel someone grab my arm. It was the dive instructor. She checked my air gauge then took my hands off the rock. I had no choice but to trust her. We drifted together for a while and I immediately began to breathe properly and clear my mask, as I was no longer fighting the force of the current. We surfaced not far from where we started.

I was glad to hit that surface and breathe air. I have never felt like that before or since. There were times on the solo trip when I was scared, terrified even, but never to the point I was that day. It didn't stop me diving that afternoon, when I did things a bit differently.

We farewelled Moira and Anika after spending five incredible nights at the Galapagos. The pair decided they would opt out of the three-week crossing to the Marquesas Islands and fly home to Australia. We said our farewells and Dave and I prepared for the 2700-mile trip. The Marquesas Islands, or Iles Marquises, were part of French Polynesia, sitting southwest of the Galapagos, smack-bang in the middle of the Pacific. I was glad Dave and I would be doing this stretch alone. I had nothing against the women, they were great company, it's just that I wanted to do a long ocean stretch in conditions as close to solo sailing as I could. Dave was a pretty mellow fellow, so I was able to pretend I was on my own much of the time.

It was a good experience, as it was the longest stretch I had done before the solo trip. I learnt how to sail a long trip in those weeks. There were heaps of things I'd spoken to John about and read that I was able to put into practice. Things like how a wind vane worked, how to set it, the fact you had to continually check it. And how you needed to constantly monitor the boat and be prepared for bad weather. But, more importantly, I learnt to take it easy and not push myself or the boat to the point where either may fail. That was the most valuable lesson I was able to carry through to my solo trip.

The most interesting lesson of those three weeks was celestial navigation. Since the moment I fell for sailing I have been enchanted by the notion of sailing under the raw power of nature, and navigating my away around the world using a sextant and the stars. To my delight, and surprise, Dave turned off the satellite navigation—the global positioning system—and navigated from Galapagos to the Marquesas Islands solely through celestial navigation. I was able to learn everything about it, and helped him plot our course daily. This reinforced my view that it was possible to rely on nature to achieve my goal of sailing around the world. I thought it was fantastic because I felt it was getting back to what sailing was all about. Each day of that stage reinforced my desire to sail around the world. I told Dave of my plans, and would run my ideas by him. For some time the idea that I should challenge David Dicks' age record had been swimming in my mind. But I really had very little sailing experience. Before I left for Belize I decided that before I set foot back in Australia I would have made up my mind whether to tackle a solo circumnavigation. That stage of the trip was pivotal to my final decision.

Dave and I really got into a different groove on that section of the trip. We'd fish and read and daydream as we sailed through mainly calm weather. Dave and I hit it off as we had similar ideals. Plus, he had sailed around the world and had so much to talk to me about, which I wanted to hear.

Meal times were my favourite part of the day. On a trip like that, you cling to anything that draws you together. For us, that was meals. They became the central part of our day. We could spend the entire day on our own, not uttering a word to each other, except to say, 'Good morning.' But when lunch rolled around, we chatted and laughed as if we hadn't seen each other for weeks. It was a real occasion. We'd set the table, put on some music and prepare the food. And it was only something that could happen with two people. When the four of us were on the boat before Galapagos, meal times didn't have the same sense of occasion, as there were always three other people to talk to. And when I was on my own I never had that sense of occasion as I actually didn't have meal times, and only ate whenever I was hungry.

But it wasn't all fun. I still had schoolwork to do as we sailed along. The curriculum was divided into weekly units. As I completed units, I'd send them back to Melbourne. I sent stuff at each stop—Panama, Galapagos and Marquesas. The final units I brought home with me from Tahiti. The work I sent in Galapagos actually failed to arrive. (I did do it, I swear I did.) It made me think that the Distance Education Centre must hear some fantastic excuses of why students' work doesn't turn up.

I was doing enough to pass the subjects, but I really hated the studying. I was sitting on a yacht on a beautiful sunny day in the middle of the Pacific reading stuff that just didn't interest me. I don't think it would have made much difference if I was in the classroom, as I would probably have been dreaming of being on a yacht on a beautiful day in the Pacific anyway.

We finally got to the magnificent Marquesas, which were extremely high, dramatic volcanic islands jutting out of the water. The islands marked the start of French Polynesia. The islands were the stuff of dreams, and finally brought meaning to the term 'heaven on earth'. Here, you can ride a horse along the bush tracks through the rainforest to huge waterfalls, with tropical fruit growing on the side of the road that you can pick and eat. The other thing I noticed was the number of tattoos on the locals. It's claimed the art of tattooing had its origins in the Pacific islands. Most people have at least one arm done with beautiful traditional designs. A few have their entire body tattooed, until they look like a moving picture. I'm afraid I fell under their spell.

Dave and I met some locals on Nuku Hiva, the main island of the Marquesas. We became friends and had dinner at their house. The woman had a tattoo that looked pretty cool. I asked where I could get one done, and the next thing I knew I had an appointment to get a tattoo. In a few days, I was sitting in the front yard of a young bloke looking through his designs in an exercise book. They looked pretty good, so I chose a turtle, which cost me $40.

There are always two parts to a story about getting a tattoo. The first is getting it done. The second is telling your mum. I didn't dare tell her while I was away. I waited until I was in the car on the way home from the airport in Melbourne, when she was still happy to see me. She didn't believe me until I actually showed her. I think she was a bit shocked. I remember her saying, 'Oh, my good little boy has a tattoo.' But she really didn't mind. I think the shock of the tattoo is wearing off because so many people have them.

We spent two-and-a-half weeks in the Marquesas, which

was just magic. It was such a social time, meeting people we'd seen in ports halfway around the world. (I'm not setting a convincing argument to then decide to sail alone and non-stop around the world, am I?) We picked up a new crewman, John, a bloke in his fifties from the Blue Mountains on the outskirts of Sydney, before pulling up anchor and heading south to the Tuamoto Islands.

The Tuamotos consist of hundreds of coral atolls that create lagoons several miles wide, which make fantastic anchorages. One of the most famous atolls in this group was Mururoa Atoll, well known as the site of nuclear weapon testing carried out by the French government. When I saw the beauty of this area, it made me angry to think that the French would do anything to harm such an environment. When I got sick from eating some of the fish we caught in this area, the locals said I had ciguatera, a form of food poisoning that had become more common since testing began.

It took us four days to reach the first atoll, Makemo, at the northern end of the islands. It was exciting to arrive at a port and see some familiar boats.

Some of my favourite memories are of the days we rowed ashore and spent the time spearing coral cod, cooking them over an open fire and eating them with damper. After lunch we'd spear some more fish for dinner, scrape the cream from some coconuts and cook it with onions and potatoes in the yacht's oven. It was a fantastic feeling, catching fish and cooking it almost immediately.

There was a lot of marine life around the atolls, and with them would come the sharks. If we didn't see more than four sharks each time we dived underwater, we'd consider ourselves unlucky. The problem with the white tip reef sharks was that when you speared a fish, you had to swim to the surface quickly and hold the struggling fish out of the water. The sharks were so aggressive they could steal from our hands, perhaps taking an arm with them.

We had a bit of fun with the sharks. Dave and I would spear a fish and then tie it with rope to a fair-sized branch broken from a tree. We'd then toss the fish back into the water, dangling it from the floating branch. A shark would swim around the fish then take it in its mouth and swim off, dragging the branch behind it. That would scare the hell out of the poor old shark, making him go faster. But no matter how fast he went, that branch kept chasing him. We'd watch from underwater and keep ourselves amused for ages. I'm sure I'll look back on this when I'm older and be horrified that I was tormenting those poor sharks.

After a week of cruising the atolls, we headed for Tahiti, perhaps the most famous of the South Pacific islands. The trip only took three days, which was way too short, as Tahiti was the end of the line for me. It was strange to be back in a city again. We pigged out on junk food and bought McDonald's sundaes. We even let our hair down and went nightclubbing. On the spur of the moment, I decided to accept a dance with one of the best-looking girls at the bar. Later that night I was told by a local that my dancing partner was actually a he. I'm glad all we did was dance!

I had decided that I had to leave Dave and his wonderful trip in Tahiti. By now I'd decided to tackle David Dicks' solo record. I'd spent many of those nights sitting on watch, staring at the stars, tossing up what to do. I think I always knew, but I needed to make it final. I called Mum from Tahiti and told her. I was going to attempt to become the youngest person to sail solo, non-stop and unassisted around the world.

The three months with Dave was a defining time in my life,

as it confirmed that sailing around the world was what I wanted to do. I hope to one day do it the same way he had, visiting fantastic places. But I had in front of me an opportunity to set out for a record. It was now or never. I organised my flight back to Australia and a week later, on 28 June, I said goodbye to Dave, thanking him for giving me the opportunity to decide what I wanted to do. I would have loved to continue to Australia, but I had to get home to prepare to leave in early December, less than six months away. The days of lazing on the boat eating fresh fish were about to become distant memories. Life was about to get pretty hectic.

CHAPTER 3

Making the Dream Come True

> Fight through ignorance, want, and care
> Through the griefs that crush the spirit;
> Push your way to a fortune fair,
> And the smiles of the world, you'll merit.
>
> Long, as a boy, for the chance to learn
> For the chance that Fate denies you;
> Win degrees where the Life-lights burn,
> And the scores will teach and advise you.
>
> — 'To My Cultured Critics', Henry Lawson

The history of round-the-world sailing is nearly as old as the art of sailing itself. The first around-the-world sailors were those led by Magellan, who in 1519 led an expedition of five ships in an attempt to sail around the globe. Although he died en route, one of those ships made it back to Spain to claim the honour.

Nearly 500 years later, sailing around the world has become one of the frontiers thousands have wanted to conquer, whether on a pleasure boat, like Dave's 40-footer, or the sleek

Making the Dream Come True

racers that glide across the outer boundaries of sailing territory in an effort to get around the globe in the quickest way possible.

Solo circumnavigation is considered the ultimate challenge. The first recorded single-handed circumnavigation was completed by Joshua Slocum, who set out from America in 1895 aboard the 34-foot *Spray*, to prove it could be done.

The most celebrated solo sailor was Sir Francis Chichester. In 1960 he won the first transatlantic sailing race in only 40 days, which was to be the precursor to his major achievement in 1966–67, when he became the first person to sail single-handedly around the world, only stopping once. He did so on the famous *Gipsy Moth IV*, which he sailed from Plymouth, in Britain, for Sydney, arriving 107 days later. It was another 119 days before he returned to Plymouth. He was knighted with Sir Francis Drake's sword upon his return. I'm not sure how many miles he did, but even though it took him 226 days, compared to my 328 days, and he pulled into port along the way, I still marvel at his achievement. He set sail without the communication gear I had, with nowhere near the safety options of *Lionheart*, with no satellite navigation and probably a pretty limited diet. I can't fathom what it must have been like for him.

It was only natural that after the first person showed it could be done, more and more people would want to do it.

The World Sailing Speed Record Council's rules state that to sail the globe, a sailor must round the five great southern Capes: Cape Horn at the tip of South America, Cape of Good Hope (South Africa), Cape Leeuwin (Western Australia), Southeast Cape (Tasmania) and Southwest Cape on New Zealand's Stewart Island. This route avoids the artificial canals of Panama and Suez. Of course, those rules would provide an unfair advantage for a sailor leaving from southern Australia. To sail the southern capes would require sailing along the 40 to 56 latitude south range and sailors would be home in no time. If the boat and sailor could handle the conditions, that is. So, to make it fair for our northern hemisphere friends, sailors from the southern hemisphere are required to enter the northern hemisphere at some stage of the journey to complete a minimum distance of 21,600 miles.

Global solo sailing is split into two camps: those who stop and those who don't. The first group consists of those who sail solo, stopping at ports as they complete the journey. Those who fall into the first group include Chichester, Tania Aebi, Robin Lee Graham and Krystina Chojnowska Liskiewicsz, a Polish woman who completed her trip on 28 March 1976 after 31,166 nautical miles and 755 days.

The second group has Sir Robin Knox-Johnson at its head. The Englishman achieved the feat in 1969 on a 32-foot ketch, taking 312 days to cover 32,000 miles. The first woman to sail solo and non-stop was Australian yachtswoman Kay Cottee, who departed Sydney on 27 November 1987, and returned to a hero's welcome on 5 June 1988. The trip took her 189 days, covering 22,100 miles.

My interest was in the age record, to become the youngest to sail around the world solo, non-stop and unassisted.

The doyen of age sailing, if you like, was Robin Lee Graham. He left the United States to sail solo around the world in 1965. What made his effort incredible was that he was only sixteen, and his boat, *The Dove*, was only 24 feet long. That is, a whole three metres shorter than *Lionheart*. Graham's voyage was incredible: he suffered all manner of misfortune, including a broken mast, was swept overboard, and narrowly escaped a collision with a tanker. The trip took him five years. He returned when he was 21, after 1739 days and 30,600 miles.

The next youngest to complete the trip was Tania Aebi,

who finished her journey on 6 November 1987, on the 26-foot *Varuna*. Her trip took two-and-a-half years; she was 21 when she returned. Hers was another incredible story. Evidently she was going nowhere fast at eighteen, so her father offered her a challenge—go to university or sail around the world. She was never able to claim the record because for 80 of the trip's 30,000 miles, she had a friend on board, which disqualified her trip as a solo voyage.

The rules, as set by the administrators of world sailing records, the World Sailing Speed Record Council, state:

> To sail around the world, a vessel must start from and return to the same point, must cross all meridians of longitude and must cross the equator. It may cross some, but not all, meridians more than once. The orthodromic track of the vessel must be at least 21,600 nautical miles in length. In calculating this distance, it is to be assumed that the vessel will sail around Antarctica in latitude 64 degrees south. A vessel starting in the Southern Hemisphere has to round an island or other fixed point in the Northern Hemisphere that will satisfy the minimum distance requirement.
>
> 'Singlehanded' means there is only one person on board. If a singlehanded skipper accepts any kind of outside assistance then the voyage is no longer 'singlehanded'.
>
> 'Without assistance' means that a vessel may not receive any kind of outside assistance whatever nor take on board any supplies, materials or equipment during a record attempt. A craft may

be anchored or beached during the record attempt but any repairs must be made entirely by the crew without outside resources or materials.

It is never permitted to take on board stores or equipment or get any other kind of help from another vessel whilst under way.

American sailor Brian Caldwell actually claimed the youngest record when he completed his round-the-world voyage in June 1996, at the age of 20 years and six months. But before he returned, someone had already set out to break that record.

On Sunday, 25 February 1996, seventeen-year-old David Dicks left Fremantle to attempt to become the youngest person to sail solo, non-stop and unassisted around the world. He had already left on his S&S 34, *Seaflight*, a week earlier to the cheers of thousands, but had to turn back after a few days, when leaking water damaged his radio equipment. Older wiser heads praised Dicks' decision. It would have taken a lot of guts to turn back. In his mind he probably imagined the doubters rubbing their hands together with glee as he re-entered port.

What made this trip unique to the other young sailors was that it was an attempt to do the trip non-stop. And he was taking an antipodal route, that is, sailing to the opposite point of departure on the globe and returning. Imagine that you stick a giant skewer through your home, until it came out the other side of the earth. That would be your antipodal point. For David, that was a point in the mid-north Atlantic, just below Bermuda. David decided, after talking with Jon Sanders, who had completed multiple circumnavigations, that he should go that extra distance to ensure there was no doubt over the record. It would also mean that his feat would equal the distance

Making the Dream Come True

of 27,000 miles that a northern hemisphere sailor must travel to round the great five southern capes. That was considerably longer than the rules required, and longer than the trip completed by Kay Cottee. She entered the northern hemisphere for her attempt, rounded St Paul's Rocks, and completed a total of 22,100 miles.

When I decided to go for David Dicks' record, there was no doubt I would take an antipodal route. The standard had been set, which I had to follow to claim his record. My antipodal point was at the Azores Islands, 1000 nautical miles off the coast of Portugal in the Atlantic Ocean.

Unfortunately for David, he was unable to claim the unassisted record, after breaking a bolt on his mast just before he rounded Cape Horn. On 24 May a Royal Navy helicopter from the Falkland Islands was summoned to drop him a replacement bolt, which was made by navy engineers. On Sunday, 17 November 1996, he sailed into Fremantle after more than 27,000 nautical miles and 264 days at sea. He was eighteen years and 41 days old.

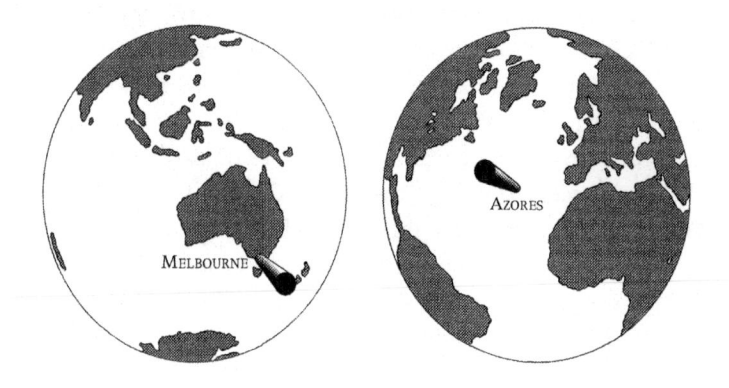

The antipodal point for Jesse's voyage.

My original plan was—and still is—to cruise the world, meet interesting people and stop in exotic ports. But I needed to raise money to do so. When the news came through that David Dicks had become the youngest person to sail the world solo and non-stop, I realised a few things.

I needed sponsorship for my adventure. To get sponsorship I needed media attention. To get media attention I had to break a record. David had set the record. So I had to beat David.

When I read about Sir Robin Knox-Johnson and his non-stop trip, I thought that sort of thing was out of my league. But when David did it at that age, I was suddenly made to realise that it was possible. All I had, however, was a plan. I had no money, no boat, no equipment and no supplies.

But I was determined to get all of those things.

~

The organisation for the solo trip was done in a ridiculously short time. Nothing had been done when I returned from Tahiti, yet I departed five months later.

I started from scratch, with no boat, no equipment, little training, and even less experience. While I hadn't mastered the finer details, such as operating electrical equipment or tweaking sails to get the best out of them, I'd learnt from Dad and Dave how to use what was at my disposal, and to take things slow and easy. I'd learnt to get a feel for what I was doing and not push things unnecessarily. Then there were the nights spent at John Hill's place practising navigation and listening to him.

When I decided to do the trip, John and I spent many hours talking over issues such as slowing the boat down when sailing downwind, lying a hull, heaving to, and survival strategies such as double harnessing, avoiding hypothermia, and what to do if the boat flipped. When I sailed from Port Phillip Bay, I was chided by many commentators for my lack of experience. What people did not know was that I had a hell of a lot of knowledge in my head.

Once the decision was made to go for David Dicks' record, I needed to get myself sorted out. The first task was to set a departure date. To become the youngest person to accomplish the record, I had to complete the trip by 6 October 1999, the day I would be the same age as David when he returned. I actually wanted to be back by my birthday, on 26 August, so I worked towards that. As David's trip took 264 days I based my planning around that, which gave me a departure of 29 November, a Sunday. Coincidentally, that was the date Kay Cottee had departed. To leave then would also reduce my chances of hitting the sort of bad weather that damaged David's equipment at Cape Horn.

I contacted the school to let them know I was not coming back. They accepted my decision, as it was obvious to them that my life was drifting in a different direction. I'd just scraped through the first semester on the work I'd sent back from overseas. Preparing for the sail was a full-time job, so I had to put school on the backburner. I thought I'd be able to complete the second half of Year 11 studies the following year, while on the trip, but that became impossible.

The real reason I enrolled and took schoolwork with me on the solo trip was in case the trip didn't come off. It placated Mum and some other members of my family.

'There is a plan in case things don't work out,' I was able to say. When I got to the halfway mark at the Azores and it came time to hand over wads of schoolwork, Mum wasn't surprised to find there was nothing to give. My only regret was that I did want to do the right thing by the Victorian Distance Education Centre, which had helped me prepare for the trip.

Lionheart

My next task was to write a proposal to potential sponsors. I was getting pretty good at proposals by this stage. I made it a bit fancy, with things like boat plans, a map of the world and details of my previous adventures. That took me a few months, because it included the budget for the trip, which had to be worked out from scratch.

How much does a solo trip around the world cost? Well, how long is a piece of string? I could be out there for 200 days or 400 days. These sorts of things were not my strong point, so it took a while, which delayed getting the proposals out. I asked for $20,000 from the major sponsor. I sent the 27-page proposal to 64 companies and organisations. The response was pretty poor.

One sponsor who came on board early was Dick Smith, founder of *Australian Geographic*. An adventurer himself, I guess he took a bit of a shine to these sorts of things. A letter arrived with a cheque for $2000. Responses like that gave me faith there would be a company willing to be my major sponsor. I was in a bit of a bind: I needed the money to get the boat, but I figured I needed the boat to show I was serious to attract some major sponsors.

I discussed with Dad and John what I should get. With a tight budget, my focus was initially on a smaller boat, the Contessa 26, the same boat used by Tania Aebi, which is known for its toughness. But I had to live on the boat, and 26 feet was not enough to store my food and supplies. I decided on something a bit bigger, making the Sparkman & Stephens 34-foot the best option. It was a classic mid-1960s design, certainly not a luxurious craft, but considered bulletproof. They were once described as the 'to hell and back' boat and boasted an impressive track record. David Dicks completed his trip in an S&S 34, and Jon Sanders had done his single and double circumnavigations in the design. It was popular not only for its strength, but the way it handled heavy weather. It had a solid keel, which weighed more than a car, and had good stability.

I can explain it the same way I explained it to Mum. Imagine a Coke bottle with a lump of lead on its side. When you put it in water, no matter how many times it rolls, it will always right itself with the lead at the bottom. The S&S 34 was also built to withstand huge pressure that comes from round-the-world sailing. The one I eventually chose turned out to be an oldie but a goodie. My worries during the trip were not with the actual boat, but with the equipment, such as my power-generating facilities, communications gear and, my biggest worry, the mast and rigging—the heart of a sailing boat. To lose them would mortally wound my voyage, and possibly put my safety at risk.

With the sponsorship proposal in circulation and the search for a suitable S&S 34 under way, I approached the major Melbourne yacht clubs to ask if I could moor at their docks and use their facilities while I prepared. I would leave and arrive at the club, which would ensure they received publicity, if the media turned out to be interested in my story.

I sent a proposal to six of the major clubs. I had replies from three, and went to see two: Sandringham and another. I went to meet with the other yacht club first. It took me ages to get there—two train trips and a long walk to the yacht club. Throughout the meeting I got the feeling they didn't think I could pull it off. That was actually said to me in the meeting in a roundabout way. I felt like an idiot sitting in front of five or six much older people more or less being told I was living in fantasy land and wasting everyone's time. It was such a negative experience. What made it worse was that it was my seventeenth birthday. The club never actually said no to me, I just never heard back from them. But, give them their dues, at least that yacht club was willing to hear my case and consider the proposal. Yacht clubs are pretty much bastions of conservatism, so a radical proposal like mine may have been beyond their comfort zone. And as it turns out, Sandringham was closer to Dad's place.

There's a modern catchphrase you hear so often—thinking outside the square. I think many people are unable to do this when confronted by a young person (albeit a slightly disorganised one) who plans to do something a bit different. They cannot imagine that they could do it, so there was no way someone else could. One motto I reckon everyone should live by is not to limit other people by their own abilities.

One thing I was not prepared for when I first started planning were the doubters. It was incredible, the number of people who were keen to disparage me and my parents. I thought that was unfair because the decision to make the trip and go for the world record was mine.

The first group were those who thought I was not going to be ready in time, and if I did, I would not complete the trip. Even in the boatyard there were people who didn't think I was going to make it. I suspect some of those actually working on the boat doubted my support crew and I were going to get things organised in time. It was just so wearing on us to have that doubt and negative vibes hanging over our heads, particularly for Dad.

We were trying to be positive and get everything done and people around us would ask why we were doing certain things. There were criticisms of the boat, of the equipment and some of our decisions. I didn't care what people thought, but when they refused to help simply because they disagreed with our view of things, that really annoyed me.

Making the Dream Come True

Still, with every goal we achieved during that preparation period came the satisfaction of knowing that we were closer to proving the doubters wrong.

Well-known solo round-the-world sailor David Adams was another who added fuel to the fire in a story in the *Australian* newspaper a week before I left. The story said, in part, 'David Adams, who sailed twice around the world in 1990 and 1994, drew breath when he heard of the attempt—Martin's first solo sail. "It's a bit like running an Olympic marathon without having run a marathon before. I admire what he's doing, but it's a big challenge."'

I didn't feel very good when I read that. As he was such a respected member of the sailing community, I was upset that it would influence many people into thinking that I couldn't achieve what I was setting out to do. But I do sometimes wonder if I was in his position whether I'd be guilty of saying the same thing.

David Adams was one of the first to congratulate me on my return. I've received many apologies since the trip. 'Listen,' they'd say, 'I am very sorry. I didn't think you would make it. You've proved me wrong, which is great.'

I even received admissions of doubt from people I didn't know. This was one letter I received when I returned:

> Dear Jesse,
> Just a short note to say how pleased and happy I am to see you safely back in Melbourne at the conclusion of your trip. I must admit that I was initially very sceptical that you would successfully complete your trip. I considered the whole enterprise ill-advised. It wasn't until you had gone about a third of the way on your journey

that I started to realise that you had every chance of success.

But the worst criticism was from those who thought I was being foolhardy. Mum and Dad never had it said to their face, but there was a lot of criticism of their role. 'How could they let their child do this?' was a common question. One man was heard to utter in front of a group of people, 'How silly was his mum for letting him go.' That was six months after I returned!

Mum had a simple view of the matter: 'I've always maintained that most people live a pretty bloody boring life. And they are the ones who come up with the adverse comments and reactions. I've never really been bothered by things that others have said. I've just gone and done what is true to me, and what I have wanted to do.'

Dad said there were 50,000 opinions out there, and I shouldn't worry about a few ill-informed ones. John Hill even copped some criticism from people he didn't know. Someone accused him of sending a kid to his death. John said they were only jealous because they didn't have the guts to do something like this.

Those people did not know me or my family. I was certainly not foolhardy. Maybe I didn't have the experience, but I made the decision because I was confident I had the ability. For people to say it was foolhardy before knowing my capabilities was quite simply measuring my abilities by their standards. I'd proven myself on other adventures, so Mum and Dad knew I had commonsense and they had confidence in me.

They also knew the effort I'd put in to get to that point. I'd been preparing for the trip for years. I may not have done any solo sailing, but I had a lot of knowledge. I knew I'd be right. I equated it to my meeting with the yacht club that didn't want

> Dear Jesse,
>   Just to wish you "God Speed," with fair winds and smooth sailing!
>   I know, from David's experience, that you won't have such ideal conditions all the time; however, when the seas are rough and the going is tough, you may like to think of a grand-mother, here in W.A., praying for your safety and success.
>   The T.V. reporters, on David's departure, asked me how I felt — my answer, "I'll pray him around the world and home again."
>   Now, I will do this for you, Jesse!
>   God bless and keep you safe!
>   Marie Hinchey.
>   David's Grand-mother

Amidst the doubters came letters of heartening support, such as this note from David Dicks' grandmother.

to help. Don't limit me by their expectations, I thought. Let me choose for myself!

Luckily I came across Steve O'Sullivan. He might not like

being described as a radical, but as the then-Commodore of Sandringham Yacht Club, he was willing to think outside the square and take a chance on me. I owe a lot to Steve, as he got the ball rolling on a lot of things.

I went to meet with Steve and the club's chief executive Scott Eccleston. I suspect they both thought I was mad at first. I told them about the trip, showed them the five-minute tape of my Papua New Guinea trip, and took them through what I required. They were very interested, and were obviously a lot more receptive to the idea than the other club. As I walked away from the meeting, I was confident they would come on board.

Steve told me he needed to take the proposal to the general committee, where he would put a strong case. And, he said with a wink, he had the final say. He told me later that the fact I'd done no solo sailing was not important to him. He could sense that I had the 'right stuff' to do it. There were others he knew who had much more experience on the water, but would never attempt this trip.

Luckily, the committee was willing to take a gamble, and a deal was struck. The club would provide me with free mooring and yard space to work on the boat, in return for the Sandringham Yacht Club name and logo on the boat. And, more importantly, I'd leave and arrive from the club. The deal was done on a handshake.

~

We travelled far and wide for a good quality S&S. Our search took Dad and me to Sydney and Adelaide, which took up a lot of precious time. The boat in Adelaide had a double spreader mast, which gives it more strength in the rigging. I was keen on this, because I was attempting to do the trip unassisted. The more strength in the boat and the equipment, the lower the chance of failure. As it turned out, *Lionheart* only had one spreader, which proved adequate. But the Adelaide boat looked a bit worse for wear and Dad wasn't too keen on it. The Sydney boat, which I'd already looked at on the way back from the Belize trip, was called *Morning Swan*. But we were worried its new paint could hide something that I wouldn't discover until I was out at sea.

We eventually found a boat in Geelong, but it was expensive. They wanted $80,000, whereas most of the boats we'd seen were $65,000 to $70,000. We were looking for a good boat, so we knew we'd have to pay a high price to get it. This boat also had good equipment, such as a wind vane and high-frequency radio. That doesn't sound like much to get excited about, but they were crucial, and not having to install them would save us valuable time. But the deciding factor was desperation—time was running out. So we bought it with money I borrowed from Mum. I ended up paying $79,900 for the boat.

We are not a wealthy family—far from it. Occasionally the money from our pamphlet rounds would help to pay the gas or electricity bill. All Mum really had was her house, but she didn't hesitate to offer to remortgage it to get the money for the boat. She was unable to remortgage it through the bank she had her loan with, as her income was too low, so she went to a solicitors' fund, which lent her the money at a higher rate of interest.

Mum best explains why she lent me the money: 'I didn't think twice about it. Some people just didn't understand that. A lot of people will have the nicest cars, but never spend money on sending their kids to a good school, because the things around them were more important. I didn't think twice about giving Jesse the money. If I was to lose the house it was not the end of the world. We could just go and rent a place.'

Lionheart

~

Lionheart was built in Perth at Swarbricks, where many of the local S&S 34s were built. It turned out to be the same age as me, although I don't think it can claim an age record for the trip. Boats built in Perth are designed to sail the Swan River, which meant the mast was built on a pivot at the deck, to enable it to lie down to get under the bridges. I believe it had done a bit of racing, but not a lot of cruising, with its longest trip up the Australian east coast.

My first impression of *Lionheart* wasn't that good. It was a dull day and, admittedly, all boats look a bit dirty on an overcast day. Most of the S&S decks I'd seen were cream, but this was grey, which made it look worse. I was probably being a bit picky, but I didn't like the dark polished timber in the cabin. But it was the extensive equipment that sold it for us. And the soundness of the boat. It may have been old, but it had obviously been treated well.

As for the story of its name, many people presume I chose the name *Lionheart* as it summed up the braveness of a boy setting out to conquer the world. It's actually less romantic than that.

My ideal name would have been *Watermark*. In fact, when Dad and I were in Sydney, the manufacturer we went to see had just started production of S&S 34s. There is a song by Enya called 'Watermark' that I love. There, in front of me, stood the most magnificent boat I had seen, a brand new S&S 34 with a dark blue hull by the name of *Watermark*. This had to be the one, I thought, until I saw the price tag—$135,000. That was that.

There had been plenty of suggestions for names. *Young Gun* was a favourite among family and friends, but seemed a bit racy for me, wankerish even. In the end, all our effort at coming up with a name was a waste, because the name of the boat when I bought it was none other than *Lionheart*. It's incredible the number of people who tell me what a fantastic name it is, and how it really captured the spirit of my adventure. But I had nothing to do with naming the boat—I actually didn't like the name to begin with. And in the rush to get ready, a bureaucratic issue pretty much forced us to keep the name. To sail out of Australian waters, a boat must be federally registered, not state registered. I was fortunate that *Lionheart* had Australian papers. To change the name and get it federally registered would have taken time we didn't have. It possibly could have taken a year to re-register, and cost $1000.

I didn't like the name because it was a bit too forward for me. It sounded like I was big-noting myself. Of course, the media started calling me '*Young Lionheart*', which was just what I didn't want. But the name started to grow on me. I now think it is a fantastic name for the boat, because it wasn't just me that did the trip—the boat also completed the voyage. So it is *Lionheart*, not me. I wasn't the one who stood up to the gales and the knockdowns.

It was a strange experience as Dad and I sailed *Lionheart* for the first time from Geelong to Sandringham, because I was the boss, not Dad. I'd just spent a couple of months on a 40-foot yacht so I was barking orders at him. We embarked on that trip with a fair degree of trepidation. We had no idea how it would sail, and if it would be a tiresome monster that would be a nightmare to handle on my own.

True to its design, it sailed beautifully. *Lionheart* was to spend only a week in the water before I left. In that time I spent a grand total of three hours sailing it on my own. That is, I did three hours of solo sailing before embarking on a trip that lasted 327 days, 12 hours and 52 minutes (or more than 7860 hours). I made some trips with Dad and other family members, but they only stretched to a few hours. My original timetable for preparation stated boldly that I'd spend a solid week sailing around Tasmania to test myself and the boat. The longest trip I did was taking a friend about 12 miles across the bay to Williamstown, where we bought fish and chips. It was not the ideal preparation, but we had no choice. If I was to leave at the end of November to break the record, the boat had to be dry-docked and prepared.

The first few weeks were simply spent inspecting every part of *Lionheart*. We had no sponsor at that stage, so Dad scurried about doing things like removing corrosion and cleaning and polishing the fittings.

The search for the major sponsor proved elusive. My first real bite was from a communications company that was just about to launch itself on the Australian market. I won't use its name as I'm not sure it got off the ground, but I was pretty excited about their interest. It got to the stage where I was to fly to Sydney to meet the head honchos. Then a call came out of the blue that they were no longer interested. It was like being on the crest of a wave that suddenly disappeared. I should have known better than to get my hopes up. It was a real kick in the guts.

But things were moving in other areas. On 21 July, the *Herald Sun* newspaper's education section, 'Learn', ran a story on how students who spend time away from school keep up with their studies. It provoked a prompt reply from the Victorian Distance Education Centre. The centre felt slighted that it had been overlooked, such was the work it carried out with sick and absent children. The email told how the centre had a student who had just returned from a trip from Belize to Tahiti—me!

The paper was aware of the interest David Dicks' trip had

aroused in Perth, particularly from schools who followed his trip. They decided to approach me about doing a similar thing, but with me writing a weekly diary to be published in the *Herald Sun*. I was taking laptop computers with me, which meant I could communicate via email. David was only able to communicate through radio, making frequent contact much harder.

A reporter from the newspaper called me and we chatted about my plans and the possibility of the *Herald Sun* becoming involved. I had intended to go to the major media outlets, particularly Melbourne's two daily newspapers, the *Herald Sun* and the *Age*, but mainly for coverage of the trip.

After the *Herald Sun* called John Hill to find out if I was fair dinkum, a deal was struck. I would write a weekly diary, and exclusive pieces for the newspaper during the trip and upon my return. The *Herald Sun* would have signage on *Lionheart* as a major sponsor and a link on my website.

At that stage, it was the search for the major sponsor that the *Herald Sun* could really help me with. It was agreed that the logo of a sponsor could be attached to the bottom of the weekly column, providing exposure to 1.5 million readers. For someone desperately trying to prove to potential sponsors that I had something to offer them, this was an arrangement straight from heaven. The other major plank of the *Herald Sun* arrangement was that they would produce an education kit free of charge to every Victorian primary school: 24 pages of information and activities based on my trip, and a map for students to follow my weekly progress. I was able to tailor much of what I wrote in the column to what was in the kits. For instance, one column focused on my daily food intake. This tied in with an activity asking students to compile a food list if they were to do a solo sailing trip. It was a good feeling knowing that I was helping kids in classrooms thousands of miles away. The *Herald Sun* also sent the kit to secondary schools as well as schools throughout Australia, New Zealand and as far afield as Austria and the United States.

The weekly columns were also placed on my website, which was promoted in the *Herald Sun*. From the website, which gave Net users the details of my adventure, students could email me. The messages went to Barbara Pesel, my sponsor's public relations consultant, who forwarded them on to the *Herald Sun*, who sent them to me to answer in my column. It was a fantastic arrangement.

I didn't realise the effect my trip was having on the schools while I was at sea. Evidently many schools based much of their work on my trip, as it covered many of their subject areas, such as maths, geography and social studies. One of my most dedicated followers was a school teacher in Austria, Franz Joseph Brändle. His class of ten- to fourteen-year-olds followed my trip through the website after he read about me in a yachting magazine. He said not many people sail in Austria because they were in the middle of Europe, and their national sports hero was a downhill skier. I wonder what they saw in me. Franz said he even saw an article in his local newspaper about my arrival home the day after I got back.

By this stage it was just over ten weeks until my scheduled departure. I now had a boat, a place to moor it and publicity with the biggest media outlet in town. It was time to do some serious hunting for a major sponsor. I re-approached some companies and spoke to new ones. But, as I was finding out, it was contacts that got you places. And contacts led me to a man called Matthew Gerard.

Steve O'Sullivan told me he had a mate who was in the electronic switch business who may be able to help out. For some reason I had the figure of $10,000 in my head. That was a big amount, and a fantastic gesture if it came true, but it was not going to solve my problems. At the time I was pinning my hopes on a phone company in Sydney. When that fell through, I went to a meeting at the yacht club where Steve introduced me to Matthew Gerard.

Matthew was head of a company called Mistral. I only knew Mistral as 'the fan company', but it was actually one of Australia's largest electrical companies, making appliances like toasters, hair dryers and kettles. It turned out they were looking for a way to promote their product range. It was a strange meeting, as Matthew didn't say much. I sensed he was sizing me up, to see if I was fair dinkum.

When he asked me what my budget was, I realised I may be onto a bit more than $10,000. I had a figure, but thought I needed a bit more, so he asked me to give him an amended budget. I raced home and re-did the budget and came up with $160,000, which did not cover the cost of buying the boat in the first place. I faxed the budget to him a few days later, then rang his secretary and organised a meeting. I was being pushy, but there was no way I was going to let this one slip through my fingers. I was pretty excited. No-one had asked for my full proposal and talked figures before.

A few days later, decked out in a suit, white shirt, a tie borrowed from John Hill and black school shoes which I actually cleaned, I caught a bus from Belgrave to Dandenong and a cab to the Mistral head office. I nervously waited in the reception area before Matthew's secretary asked me to follow her to the boardroom, which was dominated by huge windows and tropical palms. She left me on my own, so I filled the time by ensuring the proposal sat parallel with the edge of the table when Matthew rushed in. I got the impression he was dropping in on the way somewhere else.

It was all over pretty quickly. I gave him some more details of the trip and he asked some questions about the proposal.

He then said something strange. 'Well, I suppose when you get back you will be able to do some work for Mistral.'

I just said, 'Yeah, yeah.'

He also asked about school, and my likely time demands when I got back. I just thought we were chatting about things in general.

'OK, that's good then,' he said. 'I'll put you in touch with the accountant and you can send the bills to her.'

It was strange. He never said yes to me. It was as though it was assumed that he'd sponsor me. I think Steve O'Sullivan may have got in his ear and egged him into a deal. I walked out, not quite knowing what had happened. But I got the hell out of there before he changed his mind.

I rang Mum. 'I think Matthew is going to do the whole thing, but I'm not sure,' I said.

'What, pay for the whole thing?' she asked.

'Yeah, I think that was what he was saying.'

I decided to just assume that was what was happening. The next day I rang Mistral's accountant, Melina, to test the waters by asking how it would work.

I was expecting her to say, 'How would what work?' Luckily, she knew exactly what I was talking about. And it was an even bigger relief when she knew that they were kicking in for the whole lot: $160,000. Gee, it was a good feeling when she said that!

The money was to pay for equipment, labour for the boat and my supplies for the trip. They also tipped in extra money for communication costs and the expense of their public relations firm, Pesel and Carr, to handle my media commitments. As I bought equipment or used a service, I'd forward the invoice to

Mistral, which would pay it and cross the amount off the total sponsorship. It was a simple system, but it worked well.

My end of the deal was that the Mistral name would be placed on the sail, as well as prominently on the boat. There was never any compulsion to wear any of the gear emblazoned with their name, but I wanted to, as a sign of good faith and appreciation for Matthew. Besides, if they were going to buy me new gear, hey, why wouldn't I?

It was a good arrangement for both parties. The equipment I had on board was top class. I never would have had the communications gear without a company picking up my costs. I believe Mistral ended up tipping about $300,000 into the trip, including about $50,000 for my email and satellite phone bills.

Mistral got a lot of exposure out of it. It would be rare for any newspaper photo or television footage not to have the Mistral name featured prominently. I've had no complaints, only thanks. But even though Mistral was very generous, it was far from a blank cheque arrangement. I'd produced a lean budget to attract sponsorship in the first place, so I still needed plenty of help from friends and other companies to get to the starting line.

And here's the funny thing. Of the three sponsors—Mistral, the *Herald Sun* and Sandringham Yacht Club—I'd only sent a proposal to one. They were all willing to jump on board on the strength of what I told them. The *Herald Sun* actually rang me and asked to meet with me, not the other way round. All of them took a punt, which paid off for them. And I'll always be grateful for their faith in me.

CHAPTER 4

The Mad Rush

With Mistral on board, the preparation moved into top gear. It had to. I only had eight weeks until I was to leave. Dad put his building jobs on hold to become foreman of the project, and Mum juggled her work with racing about town buying supplies. Mum says those months leading up to the departure were so crazy she never had the time to consider what I was attempting to do.

There was a steady procession of tradesmen scrambling around *Lionheart* to get her ready. Phil Carr, who had taken me out on a yacht for the first time nine years before, was one of the major workers on the project. Perhaps he felt he should shoulder some of the blame for all this madness by giving me my first taste of the sea. There were a thousand things to do, some bigger than others, and so little time to do them.

One of the first jobs was to install a collision bulkhead in the boat. Basically, this meant filling the bow with foam and sealing it off from the rest of the boat. If I were to collide with something, there was a fair chance the bow would strike the object first. A collision bulkhead separated that part of the boat from the cabin area, stopping water flowing throughout

The Mad Rush

the boat and therefore reduced the chance of the boat sinking. It was not a simple process, nor was it cheap. We had to get someone to lay the fibreglass, source the foam, then install it. It also meant sealing off valuable storage space under the V-berth.

Next we turned our attention to power. We decided that I would use solar and wind power for electricity generation. Methylated spirits would be used for cooking and heating, so I took 200 litres. There was already a small solar panel on the rear of the boat, but not enough for adequate power for nine months. We decided on three 80-watt solar panels. Each panel cost $600. The first step was to design a frame for the panels, which took a lot of time. The trick with placing solar panels on a boat is that you want to catch the rays from the sun directly, as well as the reflection off the water in the morning and evening. It had to be well clear of the deck to catch all angles of the sun, but not be too far off deck to be wiped out in the first patch of rough weather.

We based our design on David Dicks' solar frame, but needed the frame to be bigger, as his 60-watt panels were substantially smaller than ours. Many yachties apparently expected the panels to be wiped clean off the boat as soon as I got out the Heads. Some even took bets on when it would occur. I have to admit to a bit of satisfaction returning with the panels still in place, having done their job without falter for 327 days. Dad deserved a pat on the back for the way he designed and had them made, even if the welders hated the hard job of installing them.

Then came the wind generator. Again, it was not a cheap instrument, costing $3500. I also took a spare generator, which I never used, and spare blades, which were required when a bird collided with the unit in the last month of the trip. You can start to see why we needed sponsors.

Installing the generator was a fiddly job with a lot of wiring. Richard, our electrician, ended up rewiring the entire boat, as well as installing a new switchboard. The wind generator turned out to be a good investment. It kicked out 30 amps in a 30-knot wind, whereas the panels would give out a maximum of 3 to 4 amps. The solar panels and wind generator were used to charge the batteries to 480 amp hours full charged. I never ran out of power, except for 24 hours off South Africa when some wires corroded.

The batteries powered my lights, CD player—which was quite often on—wind speed and wind direction instruments. My email system was permanently on but used little power except when I sent mail. The radar used quite a bit of power (2 to 3 amps an hour), but there were few parts of the trip where I would leave it on for long stretches. I initially planned to have it on permanently, but when I made contact with a fellow lone sailor I asked him if I needed to run the radar and navigation lights so far down south in the Pacific near Cape Horn.

'Don't worry, there's no-one down here,' he replied. Until that point I felt guilty having only my lights on and no radar. But he reasoned that neither would do you much good in an area where, if you were to hit something, it would be an unmanned object like a floating container, debris or an iceberg. Or another solo sailor with no lights or radar coming the other way!

All our preparations were done on the basis that the trip had to be unassisted. As David Dicks required help off Cape Horn, it was the one aspect I was determined not to fail. That meant having good equipment, good workmanship and plenty of spares. A lot of money was spent on sails, for instance. I took two new mainsails, on top of the existing one. I also took two genoas, two no. 3 jibs, two storm sails and a spinnaker.

I had to buy a new life raft, so I made sure I got a good one.

The Mad Rush

If I had to ditch *Lionheart*, I wanted at least an even chance of surviving. The one we chose was a six-man raft with a double floor to protect from the cold, and fitted with an EPIRB. It also had a survival suit with a bag of supplies to last me a few weeks.

Many people worked on the boat. Riggers took the mast out and replaced the stays, and added another inner forestay and two running back stays. These wires held the mast up, so it was important they were as strong as we could make them. A mechanic checked the motor and got it ready. A specialist calibrated the new compasses. If something wasn't new, it was taken off, fixed, cleaned up and put back on. We basically built a whole new boat. By the time we finished, I estimated we'd tipped about $100,000 in equipment and labour into the boat. That didn't count the supplies and non-attached stuff, like the communications gear.

It was a crazy time. Dad, Phil and I were flat out. Dad was working from sun-up to well into the night every day of the week. I concentrated mainly on buying supplies, coordinating other sponsors and dealing with the media. After I appeared in the *Herald Sun*, the other news organisations started to take notice, and my profile grew.

At the end of each day I'd be at the yacht club to talk to Dad and the workers. They'd take me through everything they'd done that day, and we'd talk about how I would fix it if anything untoward happened during the trip. As I was too young to hold a driver's licence, Mum had to race around picking up equipment and supplies. She was also a prolific list writer—food, clothes, cleaning products—Mum would write a list and chase it down. She also had the benefit of talking to Pat Dicks, David's mother. Mrs Dicks actually heard about me while she was in Melbourne, and came to visit us at Sassafras.

She was able to tell us what to expect, particularly from the media, and what sort of things I should take with me. Again, I was benefitting from David Dicks going before me.

Each day of preparation produced either a new drama or a goal accomplished, and everything in between. One day Dad was working in the lazarette, a storage area under the cockpit, when the lid to the access hole flipped shut, and the latch caught. He banged on the side of the boat for someone to get him out, but to no avail. Eventually he had to ring the main switch at the Yacht Club on his mobile phone for someone to help him. As they came down, someone wandered past and heard the commotion and let him out.

And there were some not so light moments. In early November we had a compass expert calibrate the new compasses. To do this, we sailed the boat in a circle while he worked everything out mathematically. As Dad motored out of the marina, the motor cut out. They started to drift towards the rocks on the shore. Dad called me on the mobile to say they were drifting toward the rocks. It was 7 a.m. and there was no-one around the yard to help us. Dad called the Water Police, who immediately responded, but reckoned they were not going to get there in time. At the last minute someone appeared in the boat yard. They grabbed a dinghy and headed out to throw Dad a line and tow the boat back in, thankfully.

Despite the frantic preparation, there was never any doubt I'd leave on 29 November. It was just a matter of what was not going to be done by the time I had to leave. As it was, most of the supplies were pretty much thrown on board. Luckily, all the major equipment had been fitted. If it hadn't, we'd have to decide whether I could mount it once I got going, or leave it behind. There were plenty of small things not done, but they were pretty minor. Really, all I needed was a pair of sails to

The Mad Rush

leave. It was very easy to get caught up in trying to make everything perfect.

As it was, we had a week's grace. My original departure date had to be put back to 6 December, after Matthew asked if that was possible. It was not due to any pressure in getting the boat ready, but so that Mistral's corporate boat could be on hand to take the media out. Their boat was not available on 29 November. I was more than willing to oblige.

People ask me if the mad rush to get everything ready ever made me stop and think, this is not how I imagined it, or that it was not worth the effort. I had no conception or expectation of doubt. Even at the lowest point, when people were doubting us, when Mum and Dad were running themselves into the ground, or when I couldn't find a major sponsor, I never stopped to reconsider the trip. I'd made up my mind, and all those experiences were merely part of the process that would lead me to my goal. My attitude was, there was work to do, so let's get into it. I never concentrated on how much work there was to do.

Amid the preparation, John Hill decided I needed to get into shape. For a few weeks, every morning at 5.45 a.m., a car would toot its horn outside my house. I'd drag myself outside into the fiercely cold winter morning of the Dandenongs, wearing my gloves and beanie. John would be waiting enthusiastically to drive me the 5 to 6 kilometres to the service station halfway down the hill. He'd keep going off to work, but I had to turn around and run home.

I hated it. The first morning, I tackled the hill with some gusto, and actually ran the entire distance. After a while I'd mix some walking with my jogging. Eventually the walking was definitely beating the running.

Looking back, I'm not sure if I needed to be in peak physical condition, to the extent where I was running up mountains at dawn. My fitness developed over the trip. Plus, I wasn't able to keep up that regime for too long, as the demands of organising the trip soon overtook me. John actually admitted to me after the trip that the fitness sessions had little to do with fitness, but were designed to get me out of bed when I didn't want to and into the cold and wet of the morning to toughen me up. Looking at it that way, I suppose it helped.

~

The three things I am asked most about are power, communications and food for the trip. The food fascinated people, particularly kids. As I did the trip unassisted, I had to take every morsel of food and drink with me.

One of the most important people to come on board in that mad preparation stage was dietitian Jacinta Oxford. I met her through a family friend. Like all of us, she had no experience with round-the-world sailing, and so spent hours researching on the Internet what other solo sailors had eaten. She then presented me with a list of foods, which I said good or yuk to. From that, she developed a menu and calculated how much food I would need for at least ten months at sea. As I could not keep anything cold, all my food had to be non-perishable, such as fruit bars, long-life dairy products (I took 270 litres of milk—a litre every day for nine months) and freeze-dried meals.

Mum, Jacinta and I went to Campbell's Cash and Carry and, armed with a large trolley, like a manual forklift to move transport pallets, we started shopping. We bought everything in one go. It cost $7400, with all of it fitting into Mum's green Mazda. It was a strange feeling, looking inside the car at every meal I was going to have for ten months. I must admit it didn't seem like enough, but looking at the daily allocations, I knew I would be right.

The Mad Rush

The daily allocations were placed in individual bags, and placed into larger weekly bags. The menu was designed in four-week blocks, so that I'd rotate my meals month by month to get a spread of nutritious food. My favourite food changed with the menu changes. When I ran out of something I'd get a new favourite.

Some things I had a love–hate relationship with. I absolutely hated freeze-dried blueberry cheesecake to begin with. I threw all the packets up the front of the boat when I came across them in my daily bags. By the end of the trip I absolutely loved it. I found myself scrambling around the storage area of the boat desperately searching for the cheesecake I'd earlier discarded. It was a bit weird. I didn't like the muesli bars, which I'd been really looking forward to, because I ate too many at the start of the trip.

My favourite meals were freeze-dried sweet and sour pork, freeze-dried spaghetti, and spaghetti in a can with meat and vegetables. Topping the list was freshly cooked damper. I'd make it in a bowl, roll it around and flatten it like a pizza, then cook it in a frying pan. Pancakes were my favourite before I discovered damper.

I think I probably would have taken the same foods without a dietitian's advice, but her advice and planning gave me a balanced diet. Without the structure, I'd probably have eaten all my favourites in one go and saved the terrible stuff for last!

I must admit, however, that I did become a bit slack when it came to the menu plans. About a month into the trip I'd open a bag, take out what I liked and throw the rest into the storage area. I lived on those 'discards' during my last month. I took food to survive ten months at sea, yet I was out there for nearly eleven months. Some people thought I would starve. Things were getting pretty desperate, but I had another few weeks of food to go, even if I was down to eating cereal without milk for lunch. Any longer and it wouldn't have been a very pleasant experience.

But it was not all freeze-dried beans and muesli. Every couple of days the bags would contain a bag of lollies about the size of my hand. Of course, I found myself getting a few days ahead of myself with the lollies. I actually lived on lollies when I departed. Minutes before I left, Jacinta gave me a huge bag of lollipops, chocolates and lollies. Combined with my collection of snakes, it kept me going for days before I got into my proper food.

My eating habits became very strange. I never ate a main meal for the day. It is amazing how much our routines are dictated by other people. The main meal for a family is a chance to sit together and talk about things. When you are on your own, there is no real reason to have a main meal. Eating merely served its base function of supplying energy. Or to fill in the time when I was bored. But I did find there were certain times when I'd eat more, usually depending upon where I was in the world. As I neared Cape Horn I'd spend most days asleep and rise at night so I could communicate with family and friends in the middle of the Australian day. And near the Azores Islands, at my halfway mark, I was up most nights doing interviews with the Australian media. So I'd eat more then, and often find myself cooking a main meal at 4 a.m.

I may have had a dietitian to advise on food, but when it came to clothes I was on my own. Pat Dicks was helpful in suggesting some of the items I'd need when I rounded the Horn—not many people realise that Cape Horn is only a few hundred miles from Antarctica. So I stocked up on the gloves, balaclavas, thermal underwear and jackets, then tackled the more mundane items.

The Mad Rush

Clothes became a strange thing for me. Take socks, for instance. My socks would naturally become dirty. Smelly, greasy and salty, to be exact, so I'd wash them in saltwater until they were less smelly and greasy, but still salty. Then I'd wear them again. But I had plenty of new unworn clean socks stowed below. More than half of them remained there for the entire trip. It sounds strange, but I didn't want to get out a new pair because then they would be spoilt.

Three weeks before I got home I did not want to bring out my spare sleeping bag, even though the one I was sleeping in was putrid. I didn't want to bring this nice clean thing out into such revolting surroundings. There was a practical side to my madness. If I got the clean clothes out, I would have no reason to wash the clothes they were replacing, so there would be more dirty clothes lying about. If I kept recycling the dirty clothes and washing them, then at least I kept the clutter down to a manageable level. Keeping my room tidy is not one of my strong points, but I found I could at least keep the cabin tidy, even though everything was salty and grotty.

My shoes didn't cause much clutter as I only took one pair. They were my skate shoes, which are like sneakers, which I'd been wearing in the boat yard the day before I left. When I got to the tropics, after not wearing shoes for months, I decided that I'd wear the skate shoes when I got off. (I didn't have many things to think about at that stage so I started to plan my arrival outfit.) I found one of the shoes, but for the life of me could not find the other. It must have fallen overboard. So, my only pair of shoes was no more. I did have sailing boots, which had to be my arriving home boots as well. At one stage I thought I'd lost one of those boots too, which would have made me look like a real idiot, getting off the boat with one sneaker and one sailing boot.

Looking back, I'm amazed at many of the things I took with me, and took home. I took ten bottles of shampoo, but didn't even use one bottle. It was pretty difficult to wash my hair, so after a while I didn't want to touch it. Instead, I tied it back. I had a few small dreadlocks by the time I returned, so it felt pretty good to have a shower at the Yacht Club and wash it.

It was the same with soap. I took a lot of it, but used very little. My first proper shower was more than two months into the trip, near Cape Horn, when I showered on the deck in the rain. Before that I'd sponge rain water over myself. Washing increased while I was in the tropics, as it was warmer. I had a solar shower, which I could hang on the mast and fill with salt water. A few times I did the whole bit—shower, shave and even dabbed on some deodorant. Pity there were no girls to impress. My return leg across the Indian Ocean was too cold to contemplate an outside wash, so it was a rare event.

The next crucial item for the trip was the electronics equipment, and the issue of how to store it.

I took three computers—I was terrified of what would happen if I only took one and it failed. I'd have been lost without a working computer, as I became so reliant on email. I took my own Texas Instruments laptop. Hewlett Packard were kind enough to donate a spare to me in exchange for a link from my website to the HP site, and the Distance Education Centre lent me the third to complete my work on.

My main worry was the effect the combination of salt and moisture would have on the internal workings of the machines. I was surprised that I ended up using my own computer until three weeks before I arrived home. I was banking on roughly three months' use out of each. That's not to say it was clear sailing with my computer, which copped a fair beating on the trip. There was a leak above the navigation table, and many

The Mad Rush

times I'd find water dripping onto the keyboard—definitely not recommended in the user's manual. I had to lick between the keys to get all the salt out before it ruined the internal workings. Occasionally it would stop as I used it. But, like most complex electronic equipment, all I had to do was give it a thump and it would fire up again. If that didn't work, I'd take out the floppy disk drive, give it a blow and fiddle about with some wires. Most times it would start the next day.

Alas, three weeks from home, the build-up of salt got too much for my technical expertise, so I switched to the Distance Education computer. Even then, the thought that I should do some schoolwork on it didn't enter my mind. (I have since had my original laptop repaired and I am writing these very words on it.)

I mainly used the laptop to send and receive emails and to write my weekly newspaper column. The satellite email system comprised a little black box, about the size of a transformer you often find on computer gear, with a small antenna in the shape of a witch's hat on the back of the boat. I'd connect the computer to the black box to send or receive a message. Given the outrageously expensive cost of satellite phone calls, email was my lifeline. The only problem I had with email came a few days after I rounded Cape Horn. For some reason, it would not work. Only after many inquiries did I discover I needed to switch from the Pacific Inmarsat-C satellite to the Atlantic satellite. It made sense, but how was I supposed to know that?

The advantage of email was that I could receive messages at any time. With the satellite phone I had to have it on to receive or make a call. It was just not possible to have it on all the time, so I had to organise a time with people if they were going to call me. Likewise with the radio. People would have to book calls in advance to get to me. It might take hours between the time the call was booked and finally speaking. But email was instantaneous. It also allowed for concise messages. Like many people, I cannot think of a single issue when I pick up a phone. I am a lot better on a keyboard when I can think about what I'm writing.

The satellite phone may have been expensive, but it was a vital tool. I approached the satellite company, Iridium, to sponsor me, but they were already committed to support an Australian trekking expedition to the South Pole. I bought the phone for $5800 and the call costs were about $15 a minute. It worked fantastically, but the costs made calls prohibitive. I was actually listening to the BBC one night when I heard a report that the satellite phone company was in financial trouble. I immediately became worried that the service would cut out on me. The strategy they adopted for survival, much to our delight, was to drop the call costs from $15 to $4.45 a minute.

My satellite phone bills averaged $2500 to $3000 a month before the price drop. One month, it got to $5000 when I hit the worst weather of the trip, near Africa, and had problems with the electronics. I found myself on the phone a lot sorting out the problems.

Sealing from moisture was vital for the electronic equipment I took: global positioning system (GPS), radar, satellite email equipment, satellite phone, three laptop computers, CD player and CDs, two video cameras, spare battery chargers, film and batteries. I stored the computers, video and spare batteries, chargers and first-aid kit in a series of pelican cases—waterproof camera cases, sealed with a rubber O-ring, with foam insulation. Not a drop of moisture penetrated the cases throughout the trip.

My main luxury on board was the CD player and more than

The Mad Rush

60 CDs. My Mistral budget had included $1500 for books, magazines and CDs, so I made sure I got what I wanted. I've been pretty influenced by my parents' music tastes as I grew up, so a peek inside my collection would reveal plenty of Bob Dylan, Van Morrison, Neil Young and Lionel Ritchie, plus more contemporary artists, such as my all-time favourite Ben Harper, Pearl Jam and Hunters and Collectors.

I found music a bit like food. I'd have a favourite for days or weeks, then I'd go right off it and find a new favourite, which I then played to death. I also took my guitar, which I was able to play a lot, particularly in the calm of the tropics. I was forever emailing Andrew, asking him to find me the chords for a certain song.

My team of advisers continued to grow. Dr Geoff Broomhall agreed to help with some medical advice. He knew a thing or two about maritime medicine, having spent sixteen years in the Royal Australian Navy. He helped me put together a first-aid kit and also came to my house one night a week for four weeks to advise me how to use the stuff in the kit. One night he brought needles and we practised the different injecting techniques on some chicken legs. We then had the chicken for tea.

Luckily there was no need to use any of his techniques, except when I stabbed myself in the finger. And, unfortunately, no chickens with sore legs ever landed on deck looking for medical treatment. In a strange way, I looked forward to the challenge if anything did happen to me. While I didn't want to break my leg, I was excited by what I'd have to do to overcome the problem.

In a perfect situation, and with more preparation time, I'd have had a psychologist on my team. All our effort went into the boat, so we weren't thinking about the actual trip. Our objective as a team was to get all the equipment ready and supplies loaded the night before I sailed. After that, I'd have to work it out for myself.

There was no denying that in the back of my mind was the fear of solitude. I don't think you can ever prepare adequately for spending nine months alone, but it perhaps would have been good to speak to someone, to talk through some issues. But it was such a race to get the boat ready, that aspect of the plan went by the wayside. Looking back, I don't think I suffered many mental crises that I was not able to work through myself.

In the same article that he wrote in the *Australian* the week before I left, David Adams made a comment that I didn't understand at the time. I do now. He said: 'I'd spent months with psychologists to prepare myself. Maybe as a kid you don't have so many hang-ups.'

I think he's right. When you are young, you have had less chance of establishing relationships that would be torn when you did something like this. There was less to pull me home mentally. But I think my biggest psychological fillip was knowing that David Dicks had achieved what I was setting out to do. And he only had two months to prepare.

Sponsors continued to sign up, right up until the week I left. There were twenty sponsors in all, from Mistral to a company that gave us some glue. Wesley, my old school, put on a drama production that raised $3000, so they had their logo on the side of the boat. That was an interesting one. I believe there was some hand-wringing at the school over whether it should be seen to support such a venture. I was, after all, virtually saying, 'Bugger school, I'm off.' Did one of Melbourne's most academically orientated schools want to send that message to other students, and risk the ire of parents who believed what I was doing was foolish? Luckily for me, they came through, and

The Mad Rush

I reckon their gesture showed that Wesley recognised that students have goals in many forms, not all of them based on classroom activities.

All those who gave me products had their names on the boat. We were given a certain brand of milk, Rev, so they got their logo on board. It seemed to appear in every picture taken of *Lionheart*. Some sponsors came on board at the last minute, under pretty strange circumstances. A few days before I left, I urgently needed some welding to attach the wind vane, the automatic steering system. A local welder agreed to do it at very short notice for $150, which was a pretty good price, so I told him I'd put his logo on the boat. He was in the right place at the right time. The electrician was a great bloke, and did a good job, so he got his sticker on as well.

When I began to plan the trip, my intention was to be the youngest person to sail solo, unassisted and non-stop around the world. But another aim to the trip arose one day unexpectedly.

Dad told me about an interview he'd heard with internationally acclaimed environmentalist David Suzuki. In the interview he revealed that a swathe of Nobel Prize-winning scientists had signed a declaration, directed at all governments, that the way air pollution was headed, air quality could deteriorate beyond safe levels within two decades. It was a bleak picture he painted, so I decided to tackle the voyage without burning fossil fuels. That is, to only use the power of the wind, through the sails, to propel me along, to harness the wind and sun for power generation, and to use only spirit fuels for cooking and heating. I figured that if I could sail the world without burning any fossil fuel, it would send a message that alternative power was feasible. I did have a diesel motor on board, but used it only as I approached the start line when I left. It was never heard again on the trip.

I contacted environmental organisations to let them know of my plans and ask if they would like to be a supporter. It was a natural extension of their work, so I was sure they'd want to come along for the ride. I contacted Greenpeace, the environmental activists. They sent me a membership form but were not interested in being involved.

I then contacted the Australian Conservation Foundation, who were keen. Or so it seemed. We met, and they agreed to get involved. My message to them was, use me! I was going on this trip and it would be a fantastic vehicle for them to get their conservation message across. I wanted them to give me information so I was armed with environmental facts to back up my arguments. It was a one-way street for them. I even had their logo prominently on the sail. All they had to do was supply me with information as I didn't have time to do the research. I wanted them to tell me what it would mean to the environment if every Australian reduced fossil-fuel consumption by a certain amount, and tips on how to cut consumption. I wanted to provide ways of fixing the problem, not just deliver the bad news. I was after the short stuff I could use in interviews. But not a lot happened.

In the end, I must admit that my trip wasn't pure in environmental terms. Even though I got around the world harnessing non-polluting energy, there was still a lot of waste. The equipment and electronics on board, the poisonous antifouling paint, and the plastic from my prepared meals were all pollutants of some description. I may not have got my environmental message across as I wanted to, but I learnt a lot from the trip. I'll make my next trip more environmentally friendly because I now know you can get around the world by harnessing the power of the sun and wind.

The Mad Rush

~

Before I knew it, Saturday, 5 December, had arrived—the day I left Sassafras to begin my journey. I would stay at Dad's that night. It was Mum's birthday, but that got lost in the hustle of preparation.

It was not the emotional time you might expect. There was no time for reflection that I was starting my trip, that I might not come back, that I was saying goodbye to the family dog, Roo. It was just another trip to the Yacht Club to get the boat ready. I never sat down with Mum and talked about leaving. There were no goodbyes with Beau or Andrew. It was all very businesslike. But that may have been a good thing. I had no time to get emotional, or contemplate the hard or dangerous times ahead. The thoughts about the trip were to come a few days after, as I sat on *Lionheart* and suddenly realised what I was doing.

But there was one special moment I'll never forget. Beau had bought a pair of tiny leather baby shoes at a secondhand shop and wrote on one of them: 'Jesse, I'm proud to be your brother. Best wishes. Beau.' One stayed on the fridge at home and the signed shoe sailed with me on *Lionheart*. The theory was that they would be reunited when I returned.

I got to the Yacht Club and did some last-minute ordering as some stuff hadn't arrived. I had to cancel one order and run to a local shop to buy wet weather gear. The jacket and pants colours didn't match, but we were beyond worrying about that.

Everything came together that day. The food came from home to the Yacht Club in a truck, where it was loaded into a shed. As *Lionheart* was still being worked on, nothing was put on board until that night when, with the help of my friends, I began the task of loading what would turn out to be my life for eleven months on board.

The sails were taken out of a trailer and onto the boat, where someone attached them to the mast and boom. The welders had only finished that afternoon and the frame they'd been working on was still a little warm. Cartons of milk were stacked along the jetty with wet weather gear, cases full of electronics and piles of other stuff. We wondered whether it would fit. The night went on and on, with only a short break for pizza. But the mood was buoyant. We all knew the amount of work needed to be done but we went on our rushed merry way with the occasional joke or laugh punctuating the packing. No-one, especially me, was thinking about the voyage ahead. We had to make it through the night first! Then we could worry about the challenge of making it around the world.

I sat back and watched as our usually disjointed family pulled together as one. Mum and Dad had gone through a rough patch while the preparations were underway. After being separated for ten years, they decided to finalise things, which meant they ended up in court wrangling over property during my preparation for the trip. It was not an easy time for anyone. So to watch them passing packages and suggesting ideas was something I really appreciated.

We worked all night. If you were to see us then, we would have looked like a bunch of zealots working with some sort of religious fervour. I certainly did not have the appearance of the cool adventurer having a good night's sleep before tackling the world.

We joked, we disagreed, we cursed and we sweated, but most of all, we got the job done.

At 7.30 a.m. we locked up the boat with 101 things still left to do. Dad and Phil dropped me around the corner at Dad's house to get an hour's sleep while they continued to St Kilda to sail Dad's 24-foot catamaran *Bohemian* back to Sandringham to

The Mad Rush

follow me out at 10 a.m. I fell on the mattress and closed my eyes. It seemed only minutes until I felt a hand on my shoulder. It was time to go. I downed an extra strong cup of coffee and had a quick shower, then jumped back into the car with Irene, Phil's wife, for the short drive to the Club. I thought about *Lionheart* as we drove along. I hadn't even laid in my bunk, yet I was about to sail her around the world.

The car rolled down Jetty Road. It was one of those Melbourne days, when the grey is punctuated by bright sunshine, squally showers and chilly wind gusts. We drove down the hill, rounded a few corners, over some speed humps and through the gate into the carpark. I grabbed an armful of bits and pieces I'd forgotten to put on board the previous night, and headed down the ramp to where *Lionheart* was moored.

Everyone was looking at me, which made me feel rather uneasy. I met Mum, Beau, Andrew and Barbara Pesel at the boat and boarded with my armful of goodies. A hand appeared from nowhere and took the items from me. They were put down below while we all spoke about how everything was going, how I was feeling and what the procedure was going to be. I noticed that I knew many of the people standing about— cousins, aunties, uncles, friends and their parents, both grandmas, and staff from the Club. I was handed small gifts by many people, the most ironic one being Jacinta's huge bag of lollies and chocolate.

My close mates had prepared a bonsai tree by cutting out little figures of each other and writing well-known quotes by that person on them. The little figures were sitting on the branches of the tree at precarious angles. I decided to leave the tree at home—I figured I had enough to worry about in a knockdown without having to think about the possibility of a bonsai plant flying into my skull.

Dad and Phil turned up in their sailing boots and overalls and got stuck into the last-minute preparation. Little ropes were tied here and there to hold things together or apart, depending on the situation, while young cousins came aboard to see what *Lionheart* looked like below deck. It was still pretty seat-of-the-pants stuff as far as getting ready went—we really went down to the wire.

As Mum, Dad and I were being interviewed by the media on the boat about twenty minutes before departure, Phil was tying the brand new sheets—the ropes that held the sails to the deck—for the first time. It was strange sitting there as the cameras and reporters crowded around. They asked me about what I expected to experience, and why I was doing it.

They asked Mum if she would miss me—'Of course.' They asked Dad how he felt. Being a man of few words, he was to the point: 'Buggered.' We all were.

I was later told that the reporters had a bit of a problem hearing our answers because a man in the berth next to me decided to start his motors, and wouldn't turn them off despite having been asked politely. Not everyone was a supporter, it seemed.

Suddenly, the time to leave arrived. If I was to make the slack water at the Heads that afternoon I had to give myself six hours to sail down there, so 10 a.m. was my strict departure time.

The flurry of preparation around me ground to a halt. Jobs had to be left undone and people started stepping off the boat onto the landing. A drum of diesel was left in the cockpit because there wasn't time to pour it into the tanks. I had to carry the fuel in case I needed to use the motor in an emergency, although I had no intention of touching the ignition of the motor the moment I got through the Heads. As it turned out, the key snapped off in the ignition during the trip, so I couldn't

The Mad Rush

start it anyway. I didn't exactly know what I was going to do with an empty drum after I'd poured the fuel into the tanks.

Time felt like it was in fast forward at that moment. Five months of frantic preparation and I was about to be cast out on my own. We hadn't tested a single thing, which increased the probability of something going wrong. And it was my first time at the helm since we'd done all the work on it. Mum gave me a teary bear hug which I blame for my occasional sore back before making her way with friends and family to *Ophelia*, the Mistral boat that was taking the media out to see me off.

I started the motor and began to organise the cockpit area. I gave a few waves and noticed one friend, Ben, standing alone in the crowd which had built to about 350. He must have missed the others, who were on board *Ophelia*. I gave him a yell to come with me. He shuffled through the crowd and jumped on board as I pushed the gear into reverse. Some men undid the lines holding me to the jetty. Apparently many people were shocked to see someone else jump aboard, as they did not realise I'd start the solo attempt at the Port Phillip Heads.

With minimal revs I drifted backwards and turned before slipping the gear into forward and gliding away over the flat water of the breakwater, thankful I hadn't collided with another craft in front of so many people and the television cameras. People were cheering and I vaguely remember a gun going off, but there was so much happening I can't be sure. I do remember passing the breakwater and into a few waves, where I began to raise the mainsail for the very first time. With the mainsail up, I cut the motor and pulled out a bit of the front sail on the furler, a system that took in and let out the genoa. The wind was from the south at about 15 knots. I was suddenly enveloped by the sounds of sailing as I cut the motor.

[105]

It was probably the quietest environment I'd been in since I got back from Tahiti.

I sat talking to Ben and waved to those aboard *Ophelia* as it cruised beside me, the television and press cameras pointing my way. Despite the attention, Ben and I spoke about normal stuff like what parties were on that weekend and how he was going to go at school the next year. The fact that I was embarking on a journey that could take the best part of a year wasn't broached. Neither was the fact we may never see each other again. I don't think I could have become emotional anyway, as I was still pumped full of strong coffee and adrenalin.

I put the thought of the trip to the back of my mind and enjoyed the sun, which was starting to peek through the overcast sky. There was also the buzz of the media and the half a dozen or so boats around us. After about half an hour the boats began heading back, and Ben and I began wondering how he'd get back to the Yacht Club. I sure as hell wasn't going to turn around to take him back, and there wasn't enough food or room for him to come around the world. I radioed *Ophelia*, but she was way too big to come alongside to collect Ben. They said another boat would be out shortly.

It was soon time for *Ophelia* to turn back. I looked over to see Mum for the last time. She was yelling, 'Put your harness on', and waving frantically. All the emotion of the past few months had overtaken her, but she was so proud of what I was doing. I know that at that moment she thought it could be the last time she ever saw me. That must have been hard for her.

Ben and I were then left alone, as we headed further and further down Port Phillip Bay. A small dot on the horizon turned out to be Dad, Phil and Ray on *Bohemian*. They'd been waylaid by the media. Ben and I watched as they battled the waves to get to us. Then suddenly, out of nowhere, a small dinghy appeared. Like

The Mad Rush

magic, this bloke had turned up to collect Ben.

'Goodbye and good luck,' we said to each other, as Ben had his own challenge ahead with VCE. With a skilful leap, Ben landed in the dinghy, and for the first time I was on my own aboard *Lionheart*. It still hadn't hit me yet.

The overcast skies parted and the sun appeared more frequently the closer we got to the Heads. Dad arrived, and we plodded on towards our goal. The starting line was drawn between Point Lonsdale and Point Nepean. The lighthouse keeper at Point Lonsdale was the official time observer who would watch me pass the line and tell the World Sailing Speed Record Council my exact starting time.

I wasn't able to find a lanyard that attached my safety harness to the boat, so Dad threw me one of his. As the afternoon wore on I spent the time opening presents and eating most of them. We got to the centre of the Bay and the markers of the West Channel. The wind was still a southerly, which caused problems when I entered the channel heading south. I was forced to tack back and forth in the narrow confines of the channels in order to make any headway. Each time I tacked I wasted time as I wasn't used to the sails and the new rig with two inner forestays. It was getting later and later, and our fear of missing the right tide started to grow. Going into the waves slowed me down considerably.

I soon realised this would be a way of life for the next nine months. To make the most efficient progress I had to point as close to the wind as possible, at about 45 degrees. I was pounding into the waves and sending spray right over the bimini, the canopy that covered the cockpit and entrance to the cabin. The excitement of the last few days and the sleepless night was catching up with me. As John Hill said, getting out of Melbourne is a bastard at the best of times, let alone when you are exhausted.

Lionheart

With a bit of water sloshing up and over the deck the leaks started to occur. I first noticed it when I went down below to get my wet weather jacket. The heat-sealed bags of food were covered in water.

One of my goals was to keep everything as dry as possible, especially my stores. And there I was, not even over the starting line, and I was bloody leaking. Field trials would have sorted this sort of problem out, I thought to myself. Through a gap in the bags I saw one to two centimetres of water sloshing across the shelf each time the boat leant to one side. This was not the start I wanted.

I radioed Dad and told him about the problem, then dropped sail and floated while I went on to the foredeck to check the anchor well, as Dad had suggested. Sure enough, it was full of water, causing an overflow that was somehow finding its way into the cabin. Dad pulled alongside and leapt from the moving *Bohemian* on board to have a look at the problem. I started clearing away the chain and anchor rope from the drain, which was obviously blocked, while Dad went below and got a length of threaded rod that was lying about, like most things in *Lionheart* that day. He poked it into the blocked escape hose and the water started to clear from the hatch. This had taken nearly an hour from the moment I discovered the water.

I was still 15 to 20 miles from the Heads. If I made a dash for it, I might have made it, but I could also have been thrown against the rocks as the water began to flow through the Heads once again. I'd missed my chance. The only safe thing to do was to drop anchor and wait until morning, or head to shore and rest overnight before striking out the next day. We decided to head to Sorrento and stay the night there. It was one of the best decisions I made on the entire trip. Ironically, it was water

The Mad Rush

leaking onto his radio that forced David Dicks to abort his first attempt to leave. Luckily, I found the fault before I passed through the Heads.

After all the previous day's rushing about and attention from the media that morning, it was a relief to finally take a break and leave when I was ready. It was a beautiful afternoon; the clouds had cleared and we made the short trip to Sorrento in only half an hour. Everyone was so exhausted. After helping me tie up to the main jetty just behind *Bohemian*, we decided to clean up and grab a bite to eat. We'd worry about drying the food in the morning. This sounded like a great idea to me. I had a second chance to really say goodbye the way I wanted, without the media looking on.

Dad, Phil, Ray and I hiked up the hill to the main street of Sorrento and ordered a special meal of mussels in a tomato sauce and salad. It felt so good to relax with some of the people who had made the trip happen and to finally share a decent meal with them the way I'd always wanted to. Every other meal we'd had the last few months was takeaway in the shadow of *Lionheart* as we sat in the boat yard.

It got dark as we unwound and chatted and ordered a second serving. Dad happened to meet a man in the restaurant who owned the local Baker's Delight that Dad had done some work for. He invited us back to his house for a shower and clean-up. Did we need it! One after the other, we took turns and appreciated the simple luxury of a shower. All I could think was, this was the last shower rose I'd see for a long time.

I was falling asleep by the time I got dressed. We got a lift back to the boat with Dad's friend. He said we could move our boats to the jetty of the Sorrento Sailing and Couta Boat Club, which had rubber edges on its poles to prevent damaging the boat. It would also provide some privacy for our work the next morning. I didn't realise, but the boat had a high profile after having been splashed all over the television news the previous night and in the papers that morning. With twenty sponsors' logos plastered across it, it was not too hard to miss, especially if we'd stayed at the public jetty. It could have been like pulling into the villages in Papua New Guinea to greet the smiling villagers, and getting stuck for hours. We had some serious work to do the next day, and couldn't afford any disturbances by well-wishers.

We thanked our new friend for his generosity. I don't think he realised how much help he'd been, and I shudder to think what I'd have been like if I'd got out the Heads that afternoon. Alas, there was not much time to consider the 'what ifs', for I think I fell asleep before my head hit the pillow.

We woke the following day and immediately tackled the last of the jobs. I was working to the same schedule of the previous day, with the aim of catching the afternoon slack water at the Heads. Pulling in for the night turned out to be a bigger blessing than we'd thought. We realised that the boom brake was still at home, with the lee bunk sheets. The sheets attach under the bunk cushions and tie to eyelets on the roof at night so the sleeping sailor doesn't roll off his bunk. It would have been a disaster if I had left without them.

Dad remembered they were still in his garage in Moorabbin. Things were looking grim. We needed someone to drive the 80 kilometres down the Peninsula. It was a Monday morning, so most people were at work, assuming I'd already left. Phil phoned his daughter's fiancé, Guy. Over a mobile phone, Dad explained where the key to the garage was and where Guy could find the bits and pieces. After several phones calls back and forth, Guy found the gear and drove to Sorrento. He found some very glad and thankful people waiting for him. What a

The Mad Rush

life saver! He stayed around and helped us finish things off.

Dad and Phil made little brackets to hold things, tied down the wind generator with an emergency lanyard and decanted the diesel into the tanks. Guy, Ray and I unloaded all the food bags to dry in the sun and set to work finding the leak from the anchor well into the boat. A small electrical wire was found to be passing through the top corner of the well via a hole a fair bit bigger in diameter than the wire.

A bit of Sikaflex (a heavy-duty silicon) on either side and some foam on the well hatch to make a better seal and things were looking pretty good. We had fresh buns and scrolls from Bakers Delight and newspapers delivered by our friend. It was funny to read stories about leaving on my trip around the world while I sat in Sorrento eating buns.

The food bags dried quickly. We used thick, black gaffer tape to seal rips in some of the bags caused by the hectic loading and unloading, then repacked them for the final time.

It was about 4.15 p.m. when the silicon guns and tool boxes were handed off *Lionheart*. Dad and the others stepped off and I followed them with handshakes and thanked them for their help. This was my second farewell, which I was able to do in my own time, the way I would have liked to have the previous day.

It was decided Dad would follow me to the Heads in *Bohemian* on his own, while Phil and Ray got a lift back home with Guy. I started up the diesel engine for the second and final time while the guys untied me from the wharf. I eased the gear into forward and pulled away from the wooden poles. Phil and Guy were waving as Dad got *Bohemian* ready.

Goodbyes were yelled: 'See you in nine months', 'good luck' and 'farewell'.

I found my way to the South Channel and started following the markers with Dad in hot pursuit. The moment of truth was fast approaching. I set the genoa and sat at the helm steering and watching the land pass by. It didn't take very long before I could see the exit to the Bay. The headlands peeled back and the opening slowly grew. My course was still about due west, towards the lighthouse, which I remembered I had to make contact with to ensure they witnessed my crossing. I whipped downstairs and got onto the VHF radio.

'Lonsdale Lighthouse, Lonsdale Lighthouse, this is sailing vessel *Lionheart*, *Lionheart*, over.'

'*Lionheart*, *Lionheart*, Lonsdale Lighthouse, go to channel six one, over.'

I changed the channel, then repeated my call. 'Lonsdale Lighthouse, this is *Lionheart*, over.'

'Yes, *Lionheart*, go ahead.'

'Oh hi, I'm just calling you up to check whether you were ready to record the time I cross the starting line? Over.'

'Roger that, *Lionheart*. We know what to do. You should be there in about 20 minutes. Good luck with it all.'

'Roger, thanks for that. Bye.'

I came out on deck and nudged the tiller a centimetre or two to starboard which would start the slow arc out the heads and into Bass Strait. I was suddenly rounding the last corner and lining up for the home straight . . . to the starting line.

I was on a starboard tack so as I came around closer to the direction from which the southerly was blowing, I winched the genoa in tighter and began getting the mainsail ready to be hoisted. It was very important to me that I cut the motor before I crossed the line to ensure I sailed around the world without using any fossil fuels. Yet I had to make sure not to cut it too early and lose manoeuvrability near any rocks or currents.

The swell from the ocean beyond the Heads was becoming more noticeable and caused the sun to disappear between great

The Mad Rush

slabs of water as they rolled by. The sun was huge and a very rich colour. It was getting closer to the horizon, lighting up the nearby coastline that guarded the bay like castles.

I looked back and saw Dad on *Bohemian* ploughing into the swell. He was travelling much faster than me, so he was doing laps around me, yelling instructions. I went forward to the mast and connected the halyard to the mainsail, to pull the sail up the mast track, and undid the ropes tying the mainsail to the boom. I looked over to the west and identified the lighthouse, then swung my head 180 degrees to the east.

At that moment my heart missed about three beats when I realised I was just about between the two heads of Port Phillip Bay. I cut the motor and, for the first time, at 5.36 p.m., I heard the sounds of my new home for the next eleven months. No gun shot to mark the start, and no spectators. Just my Dad somewhere around me on his small catamaran, a distant observer in the lighthouse, and the setting sun to witness what was the biggest moment of my short life.

CHAPTER 5

Reality Bites: Australia to New Zealand

> It was a wrenching moment. I was thinking that could be the last time I would see him.
>
> — Kon Martin

The wind slowly died after Dad turned back for Sorrento and I was left out in the cockpit steering *Lionheart*, trying to get as much distance between me and the land as possible. I felt stunted as I sat there with the sails jarring the rig as the roll from the swell lashed the whole boat.

It soon got dark, and I could still easily see the lights of land and was passed by the *Spirit of Tasmania* passenger liner on its way to Devonport. I sailed for about six hours, until I could take no more, at about midnight. I was so tired, but I had to set the radar on guard alarm. I hadn't had a chance to have a look at it before I left Melbourne, but now I was forced to. I was in a very busy shipping area, so I had to have the radar on all night. It's amazing how quickly the human mind can learn when it knows there's a point to it. In my case I knew I could go to sleep once I worked out how to set the alarm. This handy function woke the radar up every ten minutes and did fifteen sweeps of

Reality Bites: Australia to New Zealand

the horizon. If anything was detected, such as an oil tanker, it set off an alarm that would hopefully wake me. I fiddled about and it didn't take long to set it up. I crashed onto my bunk and lost consciousness.

I woke the next morning to a slight breeze that picked up as I set my course and got the wind vane, which kept the boat on course, working. What a great feeling it was to be moving comfortably and in the right direction away from land. The wind got to about 15 knots, so I put a reef in the mainsail and pulled a bit of the genoa in. I was winching the mainsail down a bit when I noticed the sail sliding straight off the mast track—another little detail we'd forgotten to attend to. I put a split pin stopper through the mast track so the sail wouldn't slide right off the mast when it was lowered. I put the slugs back into the track and raised the sail again before getting up there with the hand drill and bending the split pin once it was in place. I put the tools away and completed the reef (that is, reduced the sail area) from the cockpit with a satisfaction that made me grin from ear to ear. I knew this feeling was why I wanted to sail solo around the world. The satisfaction of self-sufficiency. Whoooo hooooo!

The wind died down again and kept picking up for short intervals only. When I was not moving I began to get very frustrated. It was a feeling that was to grow with the trip and nearly consume me at times.

I tuned into the weather reports on the VHF radio and heard that there was radio traffic being held for *Lionheart* Victor Hotel Alpha India. I went to channel 26 and called up Telstra Radio, who asked me my vessel's name and call sign.

They told me I had a call waiting from a Louise Martin. I took the call immediately and in a short while I heard Mum's voice coming through the speaker. It was so good to hear her voice. I asked about what was happening at home and she asked me how I was going. I didn't have a great deal to report other than how we stopped overnight in Sorrento before leaving and that there wasn't a great deal of wind around.

Mum, Andrew, Gran and some others had actually driven down to the Heads to watch me sail through that Sunday afternoon. When I failed to turn up they presumed I'd stopped with Dad, so they trudged off home again.

> Wednesday, December 9
> I feel worst in the morning cos everything is quiet
> and I'm salty and half asleep but during the day
> there has been heaps to do.

It was my third day at sea when it all hit me. I suppose it was inevitable, and I still can't quite put my finger on it. The afternoon sun was beaming through the portholes and bringing the teak cupboards to life in the most beautiful colour, while the boat gently rocked to and fro on the flat ocean. There was not a breath of wind. The music of Enya, combined with the beauty and the overwhelming loneliness, brought all the emotion of the previous few months to a head. I sat at the navigation table crying like a big baby, sobbing and wailing. But they weren't tears of sadness or pity—they were tears of realisation at my new life.

I'd say the frustration of not moving added to the tears, but I mainly put it down to the incredible shock. I'd always been so busy doing something back home, that to have nothing much to do left me feeling empty and frustrated. Add to this the news I was getting about my mates spending boiling hot days at the pool talking to pretty girls, and it's not surprising I felt the way I did. The crying bouts came and went—you can't physically cry 24 hours a day.

Reality Bites: Australia to New Zealand

The first few days were spent cleaning the boat, catching up on sleep and doing odd jobs while I wallowed with hardly any wind. I'd worked out the guard alarm on the radar and was getting used to the power consumption meter.

I was living on mostly lolly redskins, lollipops and chocolate. My food rations were still safely stored in the bow. When it came to start them, three days out, it was another milestone that marked the beginning of the voyage and made me feel I was really getting into the trip.

> *Thursday, December 10*
> Wind picked up a lot and stayed at about 20–25 knots for most of the day. Got two reefs and small amount of genoa out. Spoke to Megan, Ed, Mum, Andrew and Dad in the morning. Have just spotted Tasmania. There is a front coming through soon so I'll head out to try to get some sea room. I'm moving about the cabin naked cos I don't want to wear clothes as I'm constantly lying down in my bunk and don't want to get it all salty.

I was getting relatively close to land again, the west coast of Tasmania, and the predicted forecast was for a front of about 30 knots that would be blowing towards the coastline. I had to make as much distance away from there before the front came through. With the 20–25 knots of wind, the swell built up and water splashed over the foredeck. It was the first time since the anchor well problem in the Bay that I encountered wet conditions. I realised I had a problem with more leaks. The bilge constantly had water in it, and I had to turn off the bilge pump's float switch because it was being activated every 60 seconds to pump only three times. It drove me mad with its high-pitched alarm. I just had to remember to pump it out every day, before the water started seeping through the floorboards.

The leaks presented the first psychological hurdle I had to encounter before I was happy and settled. Leaks weren't life-threatening at all, and I'd have had to be unconscious for about a week before the water level rose enough to cause any serious dangers. But they were always there and I couldn't escape them. When the wind was more than 20 knots the deck joins started to trickle tears of salty water down the inside of the hull. They didn't look like much, but there were obviously enough of them to allow the water in.

I very much wanted to keep the inside of the cabin safe and dry, somewhere I could retire to when the weather got bad. However, it wasn't going to plan. First I noticed the food bags in the V-berth were once again wet, with water droplets over the plastic. Then the floor around the toilet area was constantly reflecting light from the wet surface. I was hoping it was only one of the sea cocks (taps) to the toilet leaking, and that it was merely a matter of fiddling about, turning them off and on.

One of the more annoying leaks was above the navigation table, where the companionway slide allowed water to squirt in and drip onto the table. I jammed a thin piece of foam in the gap to try to reduce the amount of water collecting on the table. That slowed down the drips, but never stopped them. The most annoying aspect of this leak was its position. The navigation table held all the electronic instruments, which were theoretically 'drip proof', but after months and months of dripping you would start to wonder. With the instruments there was the 240-volt inverter plug which powered my laptop computer and kept me in contact with the outside world through email. The navigation table was also the only place I could sit the computer when I was working on it. I was forced to keep wiping the drips from the hatch slide before they gained enough weight to fall.

The wind died down to 20 knots as the first front passed and the sun made an appearance. The silhouettes of four huge albatross circling the boat made for quite a sight.

Friday, December 11
Listened to Simon & Garfunkel twice this morning. Wind has died off to 15 knots and I am doing 4–4½ knots. Mum is trying to organise film and rubbish drop at the bottom of Tassie. Barometer has dropped from 1010 to 1005. Could be the low expected to hit tomorrow morning. Still two reefs and small amount of genoa. Can't be bothered doing much. Will make a daily timetable of routine.

TIMETABLE
08.00   Breakfast and position plot
09.00   Exercise
10.00   Radio interviews for one hour
11.00   Email for one hour
12.00   Lunch—free time
15.00   Radio interviews
16.00   Answer email
17.00   Free time
19.00   Cook dinner and clean the day's dishes

It was a nice attempt at setting some kind of routine. But it was more a writing exercise than anything practical. Plus, I don't quite know why I was setting such a tight schedule for myself. Needless to say, that, and any other routine, went out the window for the rest of the trip.

> *Friday, December 11, 6.10 p.m.*
> Gybed half an hour ago and the wind is slowly increasing to 10–15 knots. Put preventer on main cos it was slamming too much. Found some nice moisturiser that smells like Laura! [a friend] Ate two-thirds of freeze-dried stroganoff. It was yuk and I was still full from lunch. Since gybe I'm on a good heading (130°) at 5 knots boat speed increasing.

The low was meant to arrive on the Saturday morning and I thought the increase in wind was the beginning of it. But, to my frustration, it never materialised.

> *Saturday, December 12*
> Running under single-reefed main with no headsail. Doing 3–4 knots at 115°. Spoke to Mum and Andrew again this morning. I thought it was Friday until they told me otherwise. They said it was so hot there (42 degrees) that no-one was doing anything. I wish I was there skating to Ringwood pool with Trav and Ryan then going out tonight. I might give my mates a call later tonight. Have read a quarter of the *Hobbit*.
> Still have preventer on main to stop the slapping. Couldn't be bothered keeping routine but I don't find myself consciously bored even though I'm not doing much. Enya is blasting away as I

write, combined with the creaking of ropes and the sound of water passing along the hull. Even though I am remote and alone I feel unstoppable.

I was still waiting for the wind to increase, but even in the light winds I was not pushing the boat too hard. In 15 knots of wind I could have been travelling at 6 knots easily but the boat was still new to me. I planned to be over-cautious with the rig to avoid any breakages. The last thing I wanted was to have a bad breakage forcing me to jury-rig the boat (rig up a temporary mast) and limp back to land unable to finish the trip. I didn't care how long it was going to take as long as I wasn't compromising my safety or any equipment vital to the success of the trip. What made me more cautious was the fact the big blocks on one of the running backstays had come undone from a missing pin bolt. I was able to fix it with a spare part but I knew the bugs were still being ironed out and how vital it was that I didn't push things.

I was sailing along the west coast of Tasmania and beating into the waves a little. I tried to stay below as much as possible because the waves were constantly hitting the hull of *Lionheart* and sending spray across the deck. The only safe spot was behind the protection of the bimini. I ventured up to the foredeck after I finished tying up a few lines when I spotted an extra large wave ahead that looked like it was definitely going to break against the boat and drench everything. I made a quick hop, step and jump back towards the cockpit area and behind the bimini, but as I stepped on the cockpit seat my boots didn't grip and I landed with all my weight on my shin against the corner of the seat. OUCH!

I was in absolute agony for a few minutes as I closed my eyes and tried to control the pain. My whole leg ached for the rest of the day so I wrapped a bandage around it. I don't know if it really did it much good but it looked all right as I sat in bed admiring my first injury of the voyage.

The light winds continued throughout Saturday night as I approached the bottom of Tasmania, wondering what kind of mischief my mates were getting up to.

> *Sunday, December 13*
> Woke up last night and thought that Dad and Phil were on the boat fixing something so I didn't get up to check our course. This happened a couple of nights ago as well when I thought that a photographer was taking a photo and I was standing by my bunk waiting for him to finish. I ended up getting into my bunk the wrong way.

Sleep-walking was a real concern for me. I have been known to do some strange things in my sleep including the usual snoring, walking and the uncanny ability to speak Chinese perfectly. It had crossed my mind that I could accidentally fall overboard on one of my midnight tours of the boat, but there wasn't really much that I could do about it. I never actually caught myself sleep-walking but then again, who knows what I got up to.

Once I was confident in operating the email system it became the thing I looked forward to the most. I got my first proper weather report on email from Mum, who received it from Roger Badham, probably Australia's best known yachting meteorologist. Roger had previously done weather routing for the America's Cup team and advised Kay Cottee on her solo trip back in 1988, so he knew what he was talking about. I would email him with my daily position, weather conditions, course and speed, and Roger would send a four-day report each morning for Mum to pass on to me. It worked excellently for the whole voyage. The reports gave me peace of mind when I went to sleep each night. My boat wasn't fast enough to get out of the way of the incoming rough weather, but just knowing generally what to expect was a huge comfort.

Those initial days allowed me to repack a lot of gear and get used to the way *Lionheart* behaved. I soon found where to jam the opened carton of milk to prevent it from spilling and where not to leave a cup that was full. I started hanging the tea-towel through the grab rail leading out into the cockpit. Other little habits like this soon developed.

> *Sunday, December 13*
> Just spoke to fishing boat *Zylathene* on VHF. They wanted to know what a yacht was doing around this neck of the woods. Told them. They wished me luck. Lost a bucket today when leaning overboard trying to scoop up some water. Better be careful as I've only got four left.

I awoke on the morning of the 14th feeling positive and happy about how the trip was going. It was kind of overcast and I wasn't moving much, so to keep my energy and confidence levels up I whacked on the Simon & Garfunkel 'Concert in Central Park' album and sang along as loud as I could.

I kept myself busy looking for the damper mix I asked Mum to buy, and read the instructions in the cockpit about how to mix the ingredients. Time flew by when my attention was held by some kind of challenge. I really got stuck into getting the measurements right and the water the correct temperature. I kneaded the mixture for ages to get an even consistency and cut the scone-shaped rolls into exact replicas of each other. I then lit the grill under the metho stove for the first time and carefully placed the damper under the flame. I stood in the cockpit to stretch my back from all the crouching I'd been doing and looked around at the flat ocean. I wasn't moving much, but at least I had some damper on the go.

> *Monday, December 14*
> Got ropes out of the way and oiled the wind vane. Sustagen tastes disgusting! Won't drink it, will settle for lovely cool milk. Hair and clothes need a wash. Am wearing hat to get rid of hair. Feeling more confident about the trip every day even though today is dull.

The psychological effects of not moving in a forward direction were tremendous. I looked at it as wasted time. I may as well have not been eating and wasting food or reading and wasting valuable words. It made my existence seem unworthy. When I was moving, I felt needed. I had to be aware of changes in the wind direction and strength, and was always on call for whatever the boat needed, which gave me some sense of value and meaning. My mood would go up and down as the wind picked up then died again.

> *Monday, December 14, 7.05 p.m.*
> Wind has died to nothing again and I am bloody annoyed! Mainsail is slapping and doing constant unseen damage. Took main down before but the boat rolls too much in the swell so I put it back up. Fresh water tastes like it has salt in it. Will have to sort out water problems soon! Things not going the right way lately.

Reality Bites: Australia to New Zealand

The sound of a mainsail whipping against the mast and rigging with the roll of a swell is as bad as the sound of a gale to a sailor's ears. Particularly in the middle of the night as they lie in their bunk. It makes a sailor cringe every five seconds as they wait in anticipation for the shrouds to buzz with the vibrations that are slowly, but surely, causing stress fractures throughout the stainless steel rigging. This slapping sound had the same impact on my senses as hearing a fist come into contact with a face during a fight. It just isn't nice, because we all know something is being broken.

The anger that used to build up in me when there wasn't much wind nearly drove me crazy, simply because of the constant responsibility for the boat. By this I mean the decision of whether to leave the sail up or to take it down. Both ways had their disadvantages. By leaving the sail up I had a constant reminder of the damage I was doing to the rig, not knowing whether the shrouds and stays that held the mast up would handle the next gale.

The obvious alternative was to take the sail down altogether, but this had its negatives. Without the sail up, the boat would roll around in the ever-present swell. This constant rolling from side to side, which didn't happen while sailing, would dislodge small irritating things around the boat, such as a fork which would find its way into the sink and make a terrible sliding sound across the steel with each lurch of the hull. Multiply this by ten or twenty other clinking, thudding, scraping and tapping noises, and I soon had an orchestra aboard that wouldn't let up. Once the fork was put back where it came from, another small item would free itself and take its place.

Besides the noise keeping me awake, there was also the movement. Lying on my bunk, I was thrown about from side to side. If I was to sleep on my side with my face against the mattress, the motion would roll me onto my back then roll me onto my side again every five seconds. I felt like one of those clown faces at the Show with their mouths open. If the rolling was consistent then you could well imagine being able to get used to the noise and movement, but that wasn't the case. The swell would start by moving the boat only slightly, then the next wave would add to the movement. Each wave would build the sway, until the gunwales were about ten centimetres from dipping under the water line. Then, suddenly, the timing of the rolling and the waves would miss a beat and cancel all movement. The boat would almost be dead still. It was incredibly annoying. The timing between the swaying waves allowed *just* enough time for me to nod off to sleep, only for it to start all over again and wake me.

It was similar to that sensation on a train or in the classroom. I'd be on the verge of getting to sleep when my head and body would jolt. I'd find myself rolling onto my back, nearly crying in frustration at not being able to get to sleep. At least I didn't have strangers peering at me like I was a freak and having to wipe the little bit of spittle from the side of my mouth. But I couldn't hop off at the next station. I was in this for the long run. That was the hardest aspect to handle. There was just no escape from this mild form of torture.

The frustration was accentuated by having to decide whether to leave the sail up or take it down. I was able to sleep better when the sail was up (there was less rolling) but sometimes the guilt of leaving it up was too overpowering and I'd take it down. If I took the option of a good night's rest then I sure paid for it during a gale when I wondered how strong the rig was. It's kind of a rule of life—take the easy road and pay for it later. But then, it's a fine line between being fastidious and living a comfortable existence. I had no idea where that line lay, and, in many ways, that was a dark cloud that hung over the trip. Was I being disciplined enough? Only time would tell.

The other issue on my mind was a growing freshwater problem. I had two tanks, one under the seat on each side, which held about 100 litres each. They were connected by hoses to the hand pump over the sink which I used for all my fresh water. I hadn't been at sea for a fortnight and already I was pumping out brackish water from the tanks. I was seriously worried about this. Overall, I had 200 litres in the tanks and another 250 litres in plastic jerry cans. I decided to use the water from the tap for cooking only, and milk for drinking until the problem sorted itself out or I was forced to find the problem.

> *Tuesday, December 15, 7.55 p.m.*
> Just went on a huge cleaning and rearranging spree. It started when I went looking for the baby wipes and just kept going until I rearranged the whole kitchen area. I didn't find the baby wipes though! Pulled genoa in a while ago. No wind at all. Sail isn't even slapping. At least I've got that in my favour! Will get report from Roger tonight when I call Mum and Andrew. Also worked out video camera mounting on back rail and got up to chapter 11 in the *Hobbit*. Busy day!
>
> *Wednesday, December 16, 9.05 a.m.*
> Woke up to sunshine and wind. Pulled the genoa out and set my course for about 150°. Finally we are moving at 6 knots. Whooo hooooo!

Lionheart

On sunny days like that I liked nothing better than to sit out in the cockpit with something to eat and drink and just watch the boat sail along. I was still getting used to the way *Lionheart* sailed and the best ways to handle her.

The first aspect I had to get used to was tacking the boat. Because of the twin inner forestays I couldn't just turn through the eye of the wind and expect the sail to easily blow to the other side, as the forestays would prevent this happening. I had to furl the genoa up until there was almost no sail out, then tack, and when the wind was side on again I could unfurl it. The wind stayed good on the 16th and I was able to make good headway in the right direction.

Wednesday, December 16, 7.55 p.m.
There is a low about 250 miles east of Tassie with a strong wind warning. Doesn't affect me yet but I'm going to run into bad weather soon enough. Want to get it out of the way cos it is bugging me. Want to see how big the waves actually get. Above all, I'm not looking forward to a wet cabin. Will go to any length to keep it dry! Will call Dad and Mum and Andrew tonight and find out what is going on.

Wednesday, December 16, 11.05 p.m.
Got email from Mum. Roger says to stay at 47°S and no further than 150°E cos there is some very rough weather at the bottom of New Zealand over the next three days. Will keep heading south and should be at 47° in the morning then will head east. Looking forward to Chrissie presents!

With Roger's reports, I was also listening to the forecasts over the high frequency radio. I found them a bit hard to understand as the operator spoke so fast with lots of positions in latitude and longitude that went straight over my head, but I got a general idea of what was happening.

I used to leave the radio on the emergency calling channel to keep watch for any vessels in distress that were sending out Pan Pans, which were calls for assistance, or Maydays, which were reserved for the really dire stuff. Melbourne Radio would come up on this channel every so often to let people know they were doing a weather report and to change to the frequency they transmitted on. They would start with warnings about dredging vessels about the coast of Tasmania, then go into the weather report and end with the radio telephone call traffic list. Quite often there was traffic waiting for *Lionheart* Victor Hotel Alpha India. I'd feel a tingle of excitement when they called my name and immediately flick over to the Sydney Radio frequency to try to make contact.

'Sydney Radio, Sydney Radio, this is *Lionheart* Victor Hotel Alpha India.'

If the atmospheric conditions weren't very good at that particular time I'd try a few more times without reply, then I'd try another more powerful frequency until an operator came back to me.

'*Lionheart, Lionheart,* Victor Hotel Alpha India, this is Sydney Radio, How's it going there, Jesse?'

'Roger, no problems here. Wind is still pretty light but apparently there's a low heading this way, over.'

'That's a Roger, Jesse, that's a Roger. What can I do for you tonight?'

Sometimes I'd just call up to report my position for safety

reasons but more often than not it was to either receive or make a radio phone call.

'I believe there's some phone traffic waiting for me, over.'

'That's right. Two calls. One from a Mrs Raisin, who booked a couple of hours ago and the other from a Mr Gary Jamieson.'

'Roger that. I'll take the one from my grandma first, thanks.'

'OK, I'm connecting you through now.'

The next thing I'd hear would be Gran's voice.

Because the HF radio waves bounce between the outer atmosphere and the water, the time of day and the sun's effect on the atmosphere made all the difference to the radio reception. Most times I'd get through to Sydney Radio about an hour after it got dark up until a couple of hours before it got light again, but during the day I got nothing at all.

There was also quite a skill involved in calling the operators. Because there were only about four frequencies that could be used to contact them, a lot of the time the Korean fishing boats in the Pacific were on every channel talking to their wives and kids. And they tended to go on a bit, up to an hour sometimes. I had to keep surfing the frequencies to find a call I reckoned was finishing—a bit hard when the person was speaking Korean. When I sensed a conversation was coming to an end I'd get ready with my microphone in hand and as soon as the Sydney operator told the fisherman how long he had spoken and how much it cost, I'd jump in before the other boats got a chance. I'd yell out quite loudly, hoping that the operator heard me and answered my call. It all depended on the strength of my signal whether I would get to the operator. I often had to wait for several calls before I got through or try a couple of hours later.

Thursday, December 17, 5.35 p.m.

It's getting colder and colder. Every now and

Reality Bites: Australia to New Zealand

then I light the metho burner to warm up the cabin a bit. Felt a bit sick earlier on today and very tired so I took a nap. Feel a bit better now. Changed sleeping ends so I could hear radar and wind alarms at chart table better. The low east of Tassie hasn't moved yet but I think I am going to have to start heading east more. I am already at 47°23'S and 20 miles away from 150°E. Will see what Roger says tonight.

7.30 p.m.
Just heard a Mayday relay from New Zealand land station Zulu Lima Mike for a yacht sighted upturned at position 32°34'S, 171°54'E. Vessels 50 nautical miles away from area are requested to detour to look (that's not me).

8.35 p.m.
Just spoke to Mum and Andrew but Dad wasn't home. She told me about the war which has just started between America and Iraq again. There may be more shipping and armed forces around my area. I hate it when I hear there is a war! Why can't people be conscientious objectors instead of going to war?

Friday, December 18, 2.25 p.m.
Wind died again at two this morning. Still no wind. What the hell is going on? Did washing today with water caught from last night. Also worked out water pump pressure problem and took the new filter out. Now it works like normal.

Stabbed my finger in the process. Started bleeding badly. When I saw the blood I nearly fainted. Sipped water until I was with it again, then put butterfly tape on. First proper medical crisis overcome.

The next morning I got my first proper fright. I was up at 6.30 a.m. for an interview with Steve Liebmann on the Today show over the satellite phone. The battery kept running out so after I did the interview I put the phone in its waterproof case. As I was doing that, I heard a strange noise I'd never heard before. The weather wasn't very good, with a stiff wind and a bit of spray flying about. I leant back on my bunk and looked out the porthole to where the noise was coming from. At first I couldn't see anything but overcast skies. Then, as the boat rolled to one side, the base of the baby forestay swung out over the side of the boat and into view.

OH NO! The baby forestay had come apart from its join at the deck, meaning there was one less stay holding up the mast. All my paranoia about being too hard on the rig came flooding back to me as I got into some wet weather gear and harnessed up. I rushed out through the companionway after removing the washboards and clipped on the lifeline before making my way to the foredeck while running possible solutions through my head about how I'd fix the problem.

There I was, not even two weeks into the trip, and I'd pushed the boat so much that a stay had already broken. How much more disciplined did I have to be to get this baby around the world with the mast still standing? I held on to one of the side shrouds and lunged at the swinging stay as it passed by. I quickly inspected the end of it, comparing it to the plate on the deck. I then let loose a huge sigh of relief. It was only the quick release mechanism that had come undone, probably by a flicking line slowly undoing the tape holding the lever in position.

Phew! A burden lifted from my conscience. Just maybe I was actually treating the boat with the care it needed. But in the back of my mind was always the niggling thought that maybe I wasn't. I may have been relieved by what had happened, but with the onset of the overcast weather, I wasn't to stay happy for too long.

> *Monday, December 21, 10.35 a.m.*
> I went back to sleep until now. Wind has died again. From 30 to 10 knots, just like that. I am rolling like anything, and it's bloody getting on my nerves. Couldn't be bothered pulling more sail out, will wait to see if it's constant. Cabin is wet on the ground and leaks above navigation table and galley make it uncomfortable. Put burner on to keep warm. Now it's raining.
>
> *11.35 a.m.*
> I hope something happens to the boat so I can go home.
>
> *12.35 p.m.*
> Cooked mushroom pilaf. Tastes as shit as I feel! Don't feel like doing anything. Just want to sleep and dream. To dream of being away from here. There is meant to be a strong front not far from here but no sign of it. I'll be damned if I know how these weather systems work. Nothing turns out how it's meant to!

That night the wind picked up again to 25 knots and continued onto the next day. I was heading almost due north because of the wind direction and wave angle.

> *Tuesday, December 22, 12.10 p.m.*
> Not a cloud in the sky. Feel shitty again. Always feel like this in the morning. Cabin area got wet again last night. I hate it wet. I hate the whole trip so far. I just want to go home. There is nothing out here, just water and water. Feel very lethargic. Lots of dishes need doing and the cabin needs to be tidied up but it makes it hard when the cabin is wet and I couldn't be bothered. Rang Hayley last night. She's getting email soon so then we can write.
>
> *12.30 p.m.*
> Just got half knocked down. Boat came to a standstill when a wave broke onto me. No big deal.
>
> *8.10 p.m.*
> Spent most of the day in bed with a headache. Was heading north but couldn't be bothered fixing it. Got out of bed an hour ago to correct course and cook tea. Still not perfect angle (northeast) but much better. Easier on boat, backstay doesn't vibrate. Feel terrible today. Want to go home. Have been imagining myself getting rescued so I can go home! Still 25 knots and waves are getting bigger. No problem, although not very comfortable. Wonder if Mum will call tonight. She didn't last night, I called her. Email

is waiting but too wet to fiddle about with computer.

That night, just as I was stepping into bed, I had another half knockdown. A freak wave hit *Lionheart* in the wrong position, sending water squirting under the forward hatch as well as the usual places above the navigation table and galley area. It was over that quickly I didn't have time to be scared, but I really wasn't in the mood for it. The next morning, the 23rd, dawned with a nice 10-knot wind. After a good sleep I felt much better.

> *Wednesday, December 23, 6.10 p.m.*
> Weather has been good. Managed to sit outside in the sun in the cockpit for a while then cleaned up kitchen and did dishes. Cooked damper and am waiting to eat. Has turned out a nice day even though I couldn't get the computer working.

> *8.50 p.m.*
> So peaceful out here (except for Queen blasting through the speakers). Watched the sunset with a cold drink while thinking. I feel so much better today. Not sure if it can depend so much on the conditions. Roger says it will be like this for a few days. Good! Was thinking as I watched the sun go down that I'd rather be out with mates but then thought of what I'd be doing. Hoping to see Hayley but going home disappointed. Here I have such peace, sunset, calm conditions with a chill in the air that makes you feel alive. I am grateful, but not looking forward to the next front.

I hit my strongest winds of the trip so far at 30 knots. There was a bit of spray flying about and the cabin got wet, but by no stretch of the imagination was it dangerous. Though it was the strongest stuff I'd encountered to date, I wasn't overly worried by the conditions. I knew very well that I was in for much worse than a 30-knot front, but the combination of getting used to my new way of life, the homesickness and the uncomfortable conditions all played a part in the way I was feeling.

From the moment I left Port Phillip Bay I said to myself that I wasn't leaving home and getting further away from it, but rather that I was on my way home. Every mile forward was one mile closer to seeing the people I loved. It was a nice ideal, and it calmed Mum down a bit, but in practical terms it didn't work that way. All I could see on the chart was my position getting further and further away from Australia, so it was difficult to think I was actually getting closer.

In fact, the trip was a series of stages, with the completion of each a major milestone and psychological fillip. The first was the stretch from Australia to New Zealand, the next from New Zealand across the Pacific to Cape Horn. From there the next segment was to the halfway mark at the Azores Islands, followed by the stage down to the Cape of Good Hope. The final leg was from the tip of Africa to home. As much as I liked to think I was heading home, I was really heading to the end of the stage I was currently in.

Apart from the radio operators, with whom I had short conversations, and the odd journalist booking a call, the only other people I spoke to were Mum, Andrew, Dad, Gran and a guy named Gary who preferred to be called Gazbo.

When Gazbo first told me his name, I thought he said Casper, so that was what I called him until I embarrassingly realised my error. He lived in Brisbane and I heard from him every couple of weeks, right up until I lost radio contact with Sydney near Cape Horn. We'd talk for quite a while and I got to know a fair bit about him, such as he never wore shoes when flying on planes, he owned two restaurants and spent most of his time scuba-diving and sailing. He was planning a trip on his boat to Tahiti with one of his mates. Oh, and he never got out of bed until after midday. I knew the trip would be educational.

> *Thursday, December 24, 12.05 p.m.*
> Had a good sleep. Conditions are the same.
> Making good ground. Had spaghetti bolognaise
> for breakfast. Beautiful!

Email soon became the one thing I looked forward to. My friends were on holidays and every email from them took me straight back home to what they were doing. It was a real escape for me, like dreaming in my sleep. The email system, known as Satcom C, worked through satellites orbiting the earth. It gave me coverage wherever I was. I was able to send mail to any email address around the world but people sending to me had to be registered with Telstra as they would be charged about one cent per character to send a message. Only a few of my friends could afford to be registered. Those who weren't sent their emails to Barbara's office, where she took out all the spaces between the words, then sent them on to me. Mistral thankfully picked up this bill. I got those small messages about twice a week. I looked forward to them so much that I'd count the days down until the new lot was due to arrive.

I developed a habit of glancing at the little black box under the switchboard that stored the messages every time I passed the navigation table. I was looking for the little orange LED light that indicated an email had arrived.

I became so obsessed I caught myself glancing at the unit every five seconds. I'd be reading a book and peer over the top of the book several times per page. I often imagined the light was flashing when it wasn't, and had to get closer to see for sure. It was obvious I was beginning to live for email.

Satcom C Message
Subject: Monkey Boy
Date: 20 Dec
From: Paul Ryan

Hi Jesse,
Thanks for your reply! I wonder what metho-flavoured damper tastes like?
    The papers like to exaggerate and bend the facts don't they? 'Big Wave puts Teen in Agony'.
    Did banging your shin hurt as much as stubbing your toe on a cold day? That kills. (I usually let out a silent scream!) Guess wot? Davo and Darren got their names in the paper for good VCE results!
    Write back or I'll . . . Err . . . Um . . . Well, I'll think of something!

Paul

Subject: Message from Inmarsat-C mobile
Date: 21 Dec
To: Paul Ryan

Hi Paul,
At the moment there is no wind and it is driving me crazy I was just outside swearing my head off

at the (lack) of wind and decided to check for emails. Thanks for calming me down (I may have done some damage otherwise). Might give you a call over the radio.

See ya,
Jesse

Christmas day was a good day for me. There wasn't much wind and not a cloud in the sky. I sat out in the cockpit opening my presents while waiting for a conference call with Mum and a television station—it was strange wishing my family a merry Christmas on the evening news. I spoke to everyone—Mum, Andrew, Beau, Gran and all my cousins. It left me with a good feeling, not let down by the fact I was not there, as I thought I may have been.

Most of the presents were edible and there were not many of them left by Boxing Day. I also got a few strange things such as colour pencils and a toy car, which I had to store upside down to stop it sliding all over the place. Mum later told me the circumstances in which she did her Christmas shopping—at 4 a.m. the day I left—which also explained the wrapping. It was a nice calm day, so I spread out in the sun and dreamt about everyone at home stuffing themselves with food.

Boxing Day was a different matter. When *Lionheart* was sailing up the Bay, Dad noticed how low the bow was in the water. The night we stayed in Sorrento, Dad suggested that when I got some time I should move some of the weight from the front V-berth of the boat to the back to prevent my nose diving in bad weather. So I got to work moving one-litre cartons of milk from one end of the cabin to the other and storing them neatly under the sink and galley area. I approached the task with the enthusiasm of someone who had decided to dedicate their life's work to moving milk cartons.

> Saturday, December 26
> Felt better this afternoon when I was working. I now believe it is a combination of weather conditions and if I am busy or not. When I get up in the mornings I'll set myself a goal to complete that day. I'm just worried I'll run out of things to do! While cleaning found more Christmas presents!
> Wow! Christmas again! It's like finding the last Tic Tac jammed under the lid that you didn't know you had left.

I was only a day away from New Zealand and was in communication with the cray fishing vessel *Aurora*, which Mum had contacted and organised to pick up my rubbish and film. The rules of the trip permitted me to hand stuff off the boat, such as rubbish, but take nothing on board. I was allowed to take on mail at the halfway mark, but this had to be inspected by an official to ensure it contained nothing that could actually assist me in any way.

> Sunday, December 27
> Have decided not to worry about the rough weather I will no doubt encounter but rather enjoy the calm that I have now and remember the blows should pass within a day or two. Sometimes I think this is all too easy and that my imagination has built the trip into something it's not but other times I am quite apprehensive, mainly cos I haven't been in much rough weather yet!

Reality Bites: Australia to New Zealand

When I embarked on this trip, I knew I had to overcome two major hurdles. The first was crossing the start line. The moment I did that, I entered a contract with myself that I could not break—I had no choice but to finish the trip once I started it.

The second hurdle would be the weather. The strongest winds I'd ever been in, on the Belize trip, were 25 knots. I'd already hit that, so I knew I could handle rougher conditions. Although I was on my own I was confident in my ability to handle anything. But I knew I was going to get much stronger winds, and it really started to play on my mind. When would they come? How would the boat handle it? How would I react? I now know that the anticipation of an event is usually much worse than the event itself.

I was at Southwest Cape, on the southern New Zealand island of Stewart Island. Getting there had taken me longer than I expected when I set out, due to the incredibly calm weather across the Tasman. I only averaged 75 nautical miles a day, when I should have been doing more than 100 miles.

Southwest Cape was one of the five southern capes I had to round for the circumnavigation. In fact, less than a month into the trip and I'd already rounded two of the capes—Tasmania and New Zealand.

The wind characteristically died and I inched my way towards the island. I can't remember seeing coastline for the first time. It mustn't have meant that much to me. I was, however, pretty excited about the rendezvous with the fishing boat. Most of all I was looking forward to actually coming face to face with humans. I maintained radio contact with *Aurora* throughout the day.

Late that afternoon, as I swayed from side to side with no wind, I spotted a white boat on the horizon heading straight for me. I grabbed the video camera and filmed its approach. I hadn't had much to film up to then. Soon the *Aurora* was alongside. I can't remember much of the conversation as I threw them my bags of rubbish. I was offered a can of beer that I couldn't accept.

The skipper, Colin Hopkins, suggested I should have a copy of the latest weather forecast, which he had on board.

Again, I politely refused.

'Go on, you can read it, then eat it to get rid of the evidence,' he urged.

One of the crew, a big Maori man, asked in a heavy New Zealand accent, 'Where have you come from?'

'Melbourne,' I yelled back.

He had a bit of a chuckle. 'Boy, you've got a hell of a long way still to go.'

They circled the boat, taking photos of the strange boy with a long way to go, before wishing me all the best and heading off to pick up some cray pots.

I didn't like what he said, but it was true. I still had a hell of a long way to go!

CHAPTER 6

On to Everest: New Zealand to Cape Horn

I was hit by one of the worst cases of 'Sunday afternoon feeling' I'd ever experienced. You know, that feeling when the weekend is coming to an end and the dread of school or work starts to build. At home, I'd counter the depression by losing myself in the plot of a movie. Out at sea, it was not that simple. I had no company or distractions like television.

It was an incredible feeling that I blamed on seeing the *Aurora* crew. I was so excited about seeing other humans that I hadn't prepared myself for the letdown when they departed. In a way, I wished I'd never seen them. The boost I'd received did not match the depression of losing touch again.

The sun was dropping towards the horizon and everything seemed to be moving in slow motion. I wallowed in my depression, until the wind picked up and I was able to set sail again and keep moving in a southeasterly direction, keeping enough distance between land and me. I made contact with Sydney Radio that night and put a call through to Dad. The radio officer told me about the terrible time the Sydney to Hobart race fleet was having with gusts up to 80 knots, a Mayday currently in process and lots of boats out of the race. I looked out a porthole and wondered how this sedate sea could get in such a state.

I got past Stewart Island and continued for the next few days on a path between 47° and 48°S. I spent time thinking about how to handle an 80-knot blow. That was simply incredible. I later heard six people had lost their lives. Fortunately, none of the weather affecting the Sydney to Hobart race reached me.

I got over my 'Sunday afternoon downer' in time to celebrate New Year's Eve in style—pancakes, water and lolly snakes as well as a phone call to my mates who were all down at Inverloch chasing girls.

> *Friday, January 1, 12.25 a.m.*
> Have been thinking about everything that has been achieved in the last year and I find it amazing. It also excites me to think how much more will be achieved in this new year, and the many unknown adventures in the years to come.

It may have been a new year when I woke that morning, but things remained the same—my diary filled with mundane incidents that a few months ago wouldn't have warranted a moment's notice.

> *Friday, January 1, 6.20 p.m.*
> I seem to have the annoying ability to be able to hit the five millimetre thick lifelines that run along the edge of the deck all the way around the boat whenever I throw some food overboard. It's a bit like picking the slow lane at the cash register—I always manage to do it.

On to Everest: New Zealand to Cape Horn

I set a course to steer me between the various small islands scattered south of New Zealand, enjoying the best winds I'd encountered to date. Not too strong, yet constant and I was moving well.

One night, while reporting my position to Sydney Radio, the operator told me of another solo sailor who had just left Sydney a few days before. His name was Peter Keig, on the 38-foot boat *Zeal*. He was from the United Kingdom and was sailing home the same route as me. The operator was kind enough to set up a time for Peter and me to make contact.

It was the start of a friendship that was to last several months. It was good to talk to someone doing the same thing. Apart from one week when Peter was forced to stop in New Zealand after cracking one of his teeth on an unexpected hard piece of dried fruit, we stayed in contact right up until the Falkland Islands around Cape Horn.

Peter was returning home after spending time in Sydney with his girlfriend. It was the second time he had sailed UK–Australia non-stop then returned to the UK, so we talked a lot about sailing and the conditions I could expect to encounter. I can't actually recall him teaching me anything about how to get to Cape Horn or ways of sailing a yacht because, when it came down to it, you can't teach common-sense, which was really the major ingredient in completing a voyage. But just speaking to him boosted my confidence immensely.

Before coming in contact with Peter the only information I had about sailing around the world was from reading the books by those who had done it. And, to tell you the truth, they did more to scare me than inspire me to carry out my dream. But Peter was able to tell me how it was. And everything he said was reinforced by the fact I'd already been at sea for more than a month. One of the most important bits of advice he gave me was that the genuinely rough days were few and far between.

We could only speak for about twenty minutes each day, as Peter had to be more careful about his power consumption than me, but we packed a lot into those minutes. I thank Peter for being there and giving me a realistic picture of what lay ahead.

~

The more consistent winds brought with them plenty of rain clouds. I really hadn't had much rain up to that point, which meant I'd had little opportunity for a decent wash. So one night, when one of the darker clouds decided to drop its load, I grabbed the opportunity. I positioned a bucket to collect the water dripping from the end of the sail and stripped down. With a chamois in one hand and a bottle of shampoo in the other I proceeded to have a very cold fresh water shower. The spitting wasn't very heavy so I had to sponge the water from the back of a seat and squeezed it over my head. I lathered up and repeated the process to rinse it off. I got back down into the cabin invigorated and proceeded to comb my wet hair with a fork, as I'd forgotten to take a comb, before retiring to bed a clean and happy boy.

> *Monday, January 4, 9.25 p.m.*
> Whoo hooo!! Just realised that I have passed 180° longitude and am at 179°48′W! I'm now in yesterday. The dark cloud passed over and I am back to 15 knots. Am waiting for email from Mum.

I was starting to really get into my books at this stage. I'd finished the *Hobbit*, *Death of a Guru*, *Walk with Me* and was on to *Further Along the Road Less Travelled* by Scott M. Peck and *Earth Time* by David Suzuki. I found that I could read up to three books or more at a time, depending on what I felt like reading. It was great to lose myself in the words. It took me away into different worlds and gave me something to think about. I'd get into some books so much that when it came time to make a sail change I'd be on deck winching in ropes or untangling lines, but in my head I was still with the book. I'd get back in the cabin and not remember what I'd just done.

> *Wednesday, January 6, 7.45 p.m.*
> Am waiting for sunny day to dry sheet which got even more wet while hanging out to dry. Floor is all slippery from oil. Caught myself talking and giggling to myself.

A common question I'm asked was whether I talked to myself during the trip. Often I'd catch myself saying things out loud, but not to the extent of having a full-on weird conversation like 'How ya feeling today?'

'Oh good thanks, how 'bout you?'

When I did speak aloud it was more like verbalising the thoughts swimming around my head. I've been doing it since I got back. In fact, I'm doing it as I write this book. As I think of some words to write down, I say them aloud to see if they sound right and flow correctly. Everyone does it (I hope) and it was that way I caught myself speaking 'out loud' as opposed to 'to myself'.

> *Wednesday, January 6, 8.15 p.m.*
> Started reading *Dove*. Great book. I am reading four books at once now. Strange!

Lionheart

I hit a few days of fog after I passed New Zealand, which was quite an experience out at sea, especially in the dark. At night I couldn't see the horizon or the stars above me. With no moon, there was no reference point for my mind. As far as I was concerned I could be floating in space. During those foggy nights I'd turn off the cabin lights and stand at the stern of *Lionheart* just feeling her motion as she took on the waves. It was amazing. Times like that I just wanted to have someone there to experience it with me, someone to say to: 'Wow, how cool is that!'

> *Saturday, January 9, 6.40 p.m.*
> Starting to feel better now cos it's evening. Don't know why. Haven't cried for ages but just have an empty feeling. I wish I could be sharing all this with someone. Bilge still keeps filling up with water. Think the leak is somewhere near the head cocks.
>
> *9.25 p.m.*
> What a great night! First I spoke to Peter who gave me so much confidence about rounding the Horn. He also told me the BBC World Service frequencies so I've programmed them in. Excellent! Then I spoke to Mum, Andrew, Phil and Irene and am now listening to the opera on the BBC.

Getting the BBC frequencies was a cause for great excitement. It became one other thing I could do to keep myself occupied. Besides playing the guitar, reading, cooking, eating, videoing, listening to the radio, writing and receiving email, talking to

On to Everest: New Zealand to Cape Horn

Peter and sleeping, the simplest but most important pastime was dreaming.

For probably 60 per cent of the time I spent out at sea I was dreaming. I dreamt about everything.

Dreaming is my drug, the thing that brings me comfort and is a tool I've used over the years to get through the hard times. Dreaming was my way of escaping to a place I wanted to be. I always dreamed whenever I needed to escape, whether I was on a school bus on the way home or at a speech night for school.

For a large part of my life I didn't have a father figure around to take responsibility and do everything for me. Mum was trying to do that job, so I unconsciously took responsibility for many of the things a father would take on. Who knows, maybe I first wanted to take off and sail around the world by myself to get away from all that. Whatever the reason, after the Cape York trip with Dad and Beau, I'd tasted a lifestyle that I loved and gave me plenty of fuel for my dreams over the next few years.

Dreaming had led me to every adventure I'd been on. I realised that dreams can be turned into reality, a realisation that is more valuable than any material thing. Once you realise that, you become a very powerful person, because you are able to do anything. I would imagine myself in a situation I wanted to be in and visualise every aspect of that situation—every single piece of equipment or the steps needed to get there.

The Cape York trip started as a dream after Dad first spoke about doing it with his friend. I spent all of Year 10 at school dreaming about the Papua New Guinea adventure. The same with the trip with Dave on *Imajica*. I dreamt about doing a trip like that until I worked my way into a situation where it came true.

Then came the ultimate dream . . . to sail around the world.

Of course, things are only possible if dreams are acted upon.

My dreaming turned into going to every boat show for three years, grabbing brochures and deciding on the equipment. I was passionate about my dream, so I put in the effort to make it come true. I believed a kid from Sassafras, who hadn't finished school and couldn't drive a car, could get a boat, find sponsorship and ultimately make it around the world. The key to it all was passion and faith, and enough self-belief to disregard people's negative opinions because I knew I was capable of making it happen.

Getting the BBC frequencies was a high point for me, but that feeling didn't last long. The next day, I hit the kind of weather I knew I eventually had to encounter.

It was about midday and I was trying to get some sleep, without much luck. The wind strength had increased and I was kept awake by sounds of stress on the boat that I hadn't heard before. I lay in my bunk until I couldn't stand it any more and decided to get up and heave to, which meant turn the boat into the waves to ride out the weather. For some reason I only put on a jacket and boots, leaving my middle section exposed to the elements. I went outside, cringing as the cold wind blasted my face. As the waves crashed into the boat I started to winch in the furler until there was only a small piece of sail still out. Then I pushed the tiller, turning *Lionheart* nearly into the wind, and the hove to position. The wind generator hadn't been turned off and was screaming as it spun furiously. It sounded like it was trying to spin itself off the pole, swinging from side to side, out of control. If I'd put my hand up it would have taken off a finger without a doubt.

I was paranoid about getting anywhere near the thing. I waited behind the solar panels, still half naked, to see how the boat handled the new position and listened to the screaming coming from above my head. I ducked down quickly into the

On to Everest: New Zealand to Cape Horn

cabin and flicked the brake switch for the wind generator.

Nothing happened. The high-pitched scream continued. I flicked it off and on again but it made no impact on the spinning turbine. I looked up at the battery meter and saw that it was generating 30 amps of power. Holy moly!

The boat was getting knocked about quite badly, as it had swung around nearly side-on to the waves. I went outside again, holding on tight with every step I took. I could sense a knockdown easily occurring because of *Lionheart*'s position to the waves. A knockdown is classified as the boat tipping 90 degrees, with the mast lying flat on the water. I decided to deploy the drogue in an attempt to head downwind with the waves.

The drogue is a small parachute of about a metre in diameter, made of strong canvas and with webbing sewn inside. The idea was to tow it behind the boat on a long length of rope while steering the boat downwind, in the direction the waves were travelling. It had the effect of slowing the boat down and letting the waves pass underneath, which would be the most comfortable and safest position for *Lionheart* to be in. However, the danger of travelling downwind was that the boat could pick up speed on the face of a large wave and actually surf down it. If the wave was big enough, there was a real danger of the nose of the boat diving into the back of the wave in front and going head over heels or pitch-poling, as it is known to yachties. If that happened, I'd almost definitely lose my mast and crucial equipment, not to mention possibly punch a hole in the hull from the broken rigging.

I threw the drogue out the back, but I couldn't get it quite right. As I furled in the remaining sail, the drogue just drifted out to the side while the boat stayed in its vulnerable position, side-on to the waves. I fiddled with the tiller and tried to bring the boat around, but to no avail. I couldn't work out what I was doing wrong.

Something needed to be done and quickly! The waves were building and the wind strength had increased to 45 knots, the strongest I'd faced so far. I needed to get *Lionheart* streamlined with the waves and the only other way I could do that was to cast out the sea anchor. The sea anchor is a much larger parachute-shaped object of about three metres in diameter. It's designed to hang from the bow of the boat where a conventional anchor would roll from, and stop the boat dead still, then drag it in line with the waves.

The rope leading from the bow that I would tie the sea anchor to ran around the edge of the deck to the cockpit and was attached to the boat with little pull-ties. All I had to do was shackle the sea anchor to the eye of the rope and throw it out the side and the water pressure would break the thin plastic tie-downs holding the rope in place.

I took the sea anchor out of its bag, stuffed it through the lifelines and into the water, trying as best as I could not to tangle it. Sure enough, the anchor started to open and the rope tightened, popping the plastic ties. As I watched it slip under the side of the boat, my stupid mistake hit me. The rope had been set up for easy use on the starboard side, the side I naturally threw the anchor out. But this was the lee side, the protected side, and the way the boat was drifting. I should have thrown it out the port side. I was drifting onto the anchor before it had even started pulling from the bow.

I frantically ran along the deck, trying to pull the rope and anchor up again to put it on the other side, but there was no way I could drag it up. It was too far gone. I could already see the fluoro colours of the parachute starting to surface on the other side.

On to Everest: New Zealand to Cape Horn

Oh no! It was wrapped under the boat and around the back of the keel. I was stuck. I had a drogue out the back doing nothing but preventing manoeuvrability, a sea anchor caught around the keel and I was still side-on to the waves, which were reaching six to seven metres. What was I going to do?

In the heat of the moment all I could think of was to cut the line of the sea anchor before it bent the prop shaft. If this was to bend, it could crack the hull of the boat. It was a pretty desperate situation. The boat was twisting violently as the wind screamed around me. The waves continued to crash over the boat. I was exhausted and freezing.

I scrambled back to the cockpit on my knees to grab a knife, still dressed only in my bright yellow jacket and blue knee-high boots. I tried to get to the cockpit as I tangled with the two harnesses securing me to the boat. The tiller was preventing my movement, with a bucket getting in the way and a whole heap of tangled ropes reaching out at funny angles trying to trip me up.

All of a sudden, out the corner of my eye, I saw white water approaching. It was the crest of a breaking wave about eight metres high. I could feel *Lionheart* leaning over to starboard as the wave hit. All I remember was the drenching and chaos as the wave half-knocked *Lionheart* down. The boat was probably on an angle of about 55 degrees as I stood on the back of the seat across the cockpit holding on desperately so I wouldn't slip off into the water. The mast swung back to a vertical position as I scrambled out from under the bimini canopy which had until then protected the companionway hatch from flying spray. It lay over me with the frame bent back into the cockpit, blocking the entrance into the cabin.

I yelled inside my head, 'I HATE THIS!'

Most of the ropes from the cockpit were hanging over the

starboard side and my boots were full of water. I still had to get down into the cabin to get a knife so I could cut the sea anchor line. Time was pressing, for I didn't know what damage, if any, it was doing. Then the second wave hit. I was on the stern at the solar panel frame at this stage surveying the mess when I saw it coming. At least this time I was able to hold on with both arms to one of the frames holding the panels in place. Once again the boat went half over, then slowly came back to its right position again.

I had to get down below to get a knife and I had to do it quickly! I brutally yanked the frames off the bimini, folded them back out of the way and got below deck. I bashed the companionway slide with the numb palm of my hand to get it moving, then slid it right back. I gave the horizon a scan for any more big waves about to break before lifting my knees and lowering myself into the opening without taking out the washboards. The cabin was soaked from the water pouring over *Lionheart*.

I grabbed the knife and put it between my teeth. I felt like a pirate. (John Hill was to blast me afterwards for not having a knife on me at all times.) I pulled myself out of the hatch slide again, pulling the slide shut behind me, then scrambled to attach my harnesses.

I was out and had the knife to cut that blasted line. It took me longer than usual to get to the bow of the boat because of the two harnesses I was using. I was always attached to the boat by at least one of them. I got to the pulpit at the front of the boat and leant out over it to slowly saw the rope with the rusty blade. It soon gave way as I severed the remaining strands and I watched as I drifted away from my one and only sea anchor.

I looked out over the waves ahead of me and for the first

time saw the ocean in such a ferocious mood. I was awed. The waves weren't breaking all that often, but the size of the swell made *Lionheart* and me feel so insignificant.

I made my way back to the cockpit, relieved that the sea anchor had been cut, but I still had to adjust the drogue to make sure I was heading downwind. The boat headed more downwind this time as I fiddled with the steering and the amount of line I let out to the drogue. Once it was secured, there was not much more I could do. I was freezing, having been out on deck with no pants for nearly an hour, so I went down below and was greeted by a soaking cabin with books, food and undies sliding around the floor in the water.

Just what I needed to make me feel better. My dry, warm hideaway was a mess!

To the experienced sailor, it would be obvious I needed a small amount of sail up to keep *Lionheart* moving forward, so the drogue would work. I simply didn't know this. It had been a bad experience, but I'd learnt a lot from it. The gale was about a force eight on the Beaufort scale, which was officially a gale, and I'd made it through.

Besides the loss of the sea anchor and the cleaning up I had to do, I was on a mega-high. I'd never been in a proper gale before and had never used equipment like a drogue or a sea anchor before, and I'd made it through a force eight gale.

Whooo hooo! And I'd done it all on my own.

That gale wasn't as bad as I'd experience later in the trip, but the fact I was unprepared and got caught side-on to the waves made it seem worse to me. Next time, however, I knew that I would have to keep the boat moving with a little bit of sail up and that I could tow a drogue downwind.

I had made mistakes, but I would learn from them. The journey was going just how I'd planned!

Lionheart

~

It took me ages to winch the drogue in the next day—I'd let out so much line. I took the fabric from the bimini off the frame as it was torn and, if I set the bimini up again, it would have probably been ripped off again.

The days on the Pacific continued with generally better weather. I tried fixing the leaks above the navigation table and galley to no avail. I moved the anchor and all the chain from the anchor well and stored it below deck in the hope it would lighten the bow and give me more buoyancy.

The winds were consistent as I passed from the shadow of New Zealand. There were days when the slapping of the sails frustrated me and days when it was a bit too wet for my liking, but generally, I was making good progress and was happy, having made it through my first blow and settling into life on the water. There was no need for me to cry any more. I was really enjoying myself. I'd been out at sea long enough to accept this new way of life, and felt satisfied in a way I'd rarely felt before.

I was happy just sitting at the navigation table drawing plans for a log cabin that I'd one day build with my beautiful wife, or taking on the challenge of making a meal in the galley of my small boat. I really started appreciating the small things in life and enjoyed the moment for what it was. There were no school bells or weekends to look forward to, just one long process of constantly checking the boat and making adjustments 24 hours a day, punctuated by sleeping, reading, dreaming and cooking.

What I loved about this way of life was that it never got boring. I'd watch the sun reflect off the water and tinkle through the back of the wavelets, yet never get sick of admiring it, because each scene was different and original. A sunset, a

On to Everest: New Zealand to Cape Horn

menacing rain cloud, an albatross soaring through the air—no matter how many times I saw those scenes, they always gave me something different to think about. Plus, they were such beautiful sights, I always felt contented and lucky to have seen them.

One night, while crossing the mid-Pacific, I felt a little uneasy about the weather and the potential for it to get worse. I went outside in my wet weather gear to monitor the sail set-up. I looked up and through a clearing in the overcast sky I saw a few stars shining so bright and clear.

I realised then that it was just the cloud cover giving me the strong winds and that above the clouds was a beautiful sea of stars stretching across the universe. I felt so relieved, because way out there, it was calm and clear and under control, compared to the conditions I was battling. I knew that the creator of it all was on my side and what seemed to be chaos for me was merely a small covering of cloud, ultimately controlled by the very being who was looking after me. What reason did I have to be concerned?

∼

As I said, even the most simple and mundane tasks on board took on a major significance. Everything I did was the most important thing in the world at that time. The process of making a chocolate milk, for instance, became not just making a drink, but a major leisure activity.

First, one tablespoon of drinking chocolate was put into a mug, then one tablespoon of sugar added. The jars of chocolate and sugar were promptly put back in their place on the shelf. I heated about 20 ml of water over the stove, which was then added to the mug until the sugar and chocolate dissolved. Then came the most important job—getting a smooth texture. The little globules of chocolate sticking together were squashed with the back of a spoon against the side of the cup. This could take a few minutes while I was lost in thought looking out the porthole. Once the consistency was satisfactory it was time to add the milk. Once poured in and stirred, I not only had some perfect chocolate milk, but I'd kept myself busy for ten minutes.

It was the same with cooking, cleaning fish, cleaning the cabin or washing dishes. Each little job became an important mission. I'd settled into a comfortable groove, but my up and down days continued, in sync with how I progressed.

> Saturday, January 16, 7.25 p.m.
> Haven't done much at all today. In fact haven't done anything. It is as if I am brain dead.
> Batteries are about negative 60. Won't run tri-colour tonight. I brushed my teeth. Felt good. Should do it more often.

I was a little concerned at what seemed to be a minor memory problem. My short-term memory seemed to have disappeared. It was terrible. I'd decide to make a hot drink for something to do, so I'd light a match and turn on the metho stove to find that the metho tank had run out. Not a problem. I'd put out the match, grab the bottle I used for transporting the metho, get the tap handle from the galley drawer and fill up the bottle from the main tank underneath my bunk. I'd put the tap handle back and squirt the metho into the stove tanks then screw the lid back on and wipe up the spilt metho. Easy! Then I'd stand looking at the stove and wonder what the hell I was going to do next. I couldn't remember for the life of me why I was lighting the stove in the first place. Ten minutes later and a few pages further into a book I'd feel like having a hot drink. It would then hit me that was what I'd planned to do in the first place.

On to Everest: New Zealand to Cape Horn

When I decided to do the solo voyage, one of the things I was looking forward to was to put into practice the celestial navigation I'd learnt from Dave on the Belize trip. I had a global positioning system (GPS) on board, which provided a pinpoint accurate position reading, but I wanted to use the traditional method as often as I could. As I got more familiar with the boat and my passage, I decided to get out the sextant, only to discover I'd left a vital piece of the process at home—the almanac (nautical tables) that I needed to do the calculations to provide a position. Without it, the sextant was as much use as a fishing rod without a line or hook. I didn't know whether to laugh or cry. I had no choice but to continue with the GPS, which took a fix from satellites circling the globe, then plot my longitude and latitude on my chart.

The excitement really started to build as I made my way further across the Pacific, closing down on the Horn. I was told by Andrew, who was religiously plotting my daily position on a huge world map at home, that my course started to take on a more assertive look. Rather than wandering all over the place, my daily positions reflected a more focused sailor who went the extra mile to head in the right direction. I was doing much better distances each day, more than 100 miles a day consistently.

Since I've been back and seen the map marking my daily positions, I don't think it's much coincidence that when the change in my progress took place, I was two-thirds of the way to the Horn. That was when my attitude and confidence surged.

I was getting pumped up as the notorious Cape Horn started to show its ugly head on the other side of my chart and my discussions with Peter sheeted home the reality that I was really going to round the Horn. Everyone goes on about Cape Horn, but it really is the Mount Everest of sailing, because of its dangerously high latitude and fierce history. If all I did was round Cape Horn on that trip, I'd have been happy. And I was going to do it all on my own.

I reckon anyone can inspire themselves. For me, the knowledge that I was soon going to round Cape Horn on my own was enough to inspire me to keep going. It was an amazing thought. Here I was, doing it on my own.

ON MY OWN!

This is what I loved about the trip. I was (to a degree) in control of my own destiny. If I was tired or lazy and didn't pull out a reef to increase my speed then it would take longer to get home and see those I missed. If I was too blasé about the maintenance and use of my equipment, then my trip could suddenly end. My decisions could affect my survival. I'd never carried such heavy responsibility before. It made me feel alive.

Before I set out on the trip, rounding the Horn and Southern Ocean sailing were such daunting challenges. But by that point I felt differently. I'd taken the plunge and faced my fears by crossing the starting line. I had faith in *Lionheart* and, after tackling the force eight gale, I was beginning to understand how the ocean worked, that the big waves just went rolling under me 98 per cent of the time without a hint of breaking into the boat. And I'd discovered that the unknown was worse when it remained unknown. When the truth became apparent, everything was so much easier. I was pumped and buzzed with excitement at my new understanding of so many things.

When I began the trip, I thought I had to follow Kay Cottee's and David Dicks' routes because they knew what they were doing. But I had become one of them. I was confident in my knowledge of the characteristics of the ocean, so I took my own route. They went this way and that way—well, I'll go my own way!

On to Everest: New Zealand to Cape Horn

The crazy thing about this confidence thing is that it is not secret knowledge you learn by sitting around a yacht club bar with a beer in hand for 50 years. It comes from going out and taking full responsibility for oneself.

I acknowledge that some experience was necessary. Many people said my three-and-a-half months of offshore sailing was not enough, but they missed the point. You can crew on a boat for most of your life but if you're just sailing in the Bay under the command of a skipper then they are the only conditions in which you will excel.

I wanted to become a good solo sailor so I could sail around the world. I had the time with Dave aboard *Imajica* where I learnt to operate the wind vane and to do celestial navigation. I was confident I could physically do everything that Dave did.

My next practice was on this trip. How can one get solo sailing experience without getting out there on their own? How do skateboarders pull off the most amazing moves mid-air without attempting the very first one?

The point is, they can't. Their strength is, they're not afraid to fail in the name of giving it a go. The skateboarder hits the tarmac over and over until he discovers that it doesn't hurt quite as much as he imagined it would. He gains confidence and learns from what he did wrong last time. I go to skate competitions and watch those blokes pulling their moves and wonder how they do it. It's the celebration of the skater's courage at giving it a go in the first place. That's the human spirit shining through. I'd come through the first blow and knew what the tarmac felt like—it wasn't so bad. It was a liberating feeling.

∼

On 17 February I received an email from Roger telling me the current weather patterns would give me a good chance to start my descent to the Horn. Until then I'd stuck to about 43°S. Cape Horn was at 56°S so I had a fair way south to go, which meant a greater chance of coming across bad weather and much colder conditions.

It didn't faze me much, though. Peter had caught up after his stop in New Zealand and was the same distance from the Horn as I was but five degrees or 300 miles further south. It was the closest I ever got to my new friend.

I got the storm jib out and hanked it permanently to the baby forestay, ready to raise at short notice and set *Lionheart* on a southeast course.

It may still have been summer in the southern hemisphere, but it made little difference once into the high latitudes. The mornings started to become very cold. I lay in bed in my thermal undies with my head under the sleeping bag breathing onto my body. I'd get up and get straight into my wet weather trousers and polar fleece jacket and boots, then light the stove to warm up the cabin.

Through my radio contact with Stewie Howarth at the Sandringham Yacht Club I made contact with Neil Hunter aboard *Paladin II*, the only Australian competing in the Around Alone Race. He was also sailing under the Sandringham Yacht Club colours. Neil's boat was a 40 footer and couldn't really compete against the state-of-the-art 60 footers in the lead. But there were quite a few boats in Neil's situation and this group would keep a radio sked twice a day. (A sked is a yachtie term for a scheduled radio link-up.)

There was Neal Peterson from South Africa on *No Barriers* and Minoru Saito from Japan who, at 65, was on track to become the oldest solo sailor. There was also a Russian adventurer, a little way behind. The race winners used to get $10,000, a pittance compared to what they outlaid, but with the

On to Everest: New Zealand to Cape Horn

race having few funds, each of the competitors in the 1998–99 race were given a T-shirt with the words 'Around Alone Veteran' on it. That was all! There was no prize for the winner.

I joined the radio sked for this group, which was about three weeks behind me but closing in fast. One night I was amused to hear from Neil that the Russian sailor got the fright of his life when sailing with a pod of dolphins. Apparently the boat broached a little on a wave just as a dolphin was jumping out of the same wave causing it to land in the cockpit. I could just imagine a big dolphin flapping about in my cockpit and didn't envy the job of getting one back into the water.

∼

Since it was first discovered Cape Horn has been known as one of the most treacherous capes in the world to round. A major reason for this is its proximity to Antarctica, about 500 miles from the Antarctic convergence zone. Being so far south compared to the Cape of Good Hope and the other capes, the frequency of gales is much higher. Add to this the chance of colliding with an iceberg, and the bitterly cold temperatures, and it's easy to see why it has its reputation.

For the two or three weeks as I headed south, the chance of bad weather increased dramatically. The main reason for that was the low pressure systems that tear around the Southern Ocean in an easterly direction between 40° and 60°S. The winds pass under all the capes of the world until they get to Cape Horn, where the shape of the land pushes the systems further south. This change of course can push the systems into awkward positions and even into each other, producing the violent weather the area was renowned for. On top of that, the water got shallower closer to land, making for larger and messier waves.

I continued moving along in good weather. I say good weather relative to what I knew the area could produce. It was still strong but not so strong that I was worried. I did notice how much the swell built up, even with little wind. It was mostly overcast weather with very few days where the sun could be seen.

The leaders of the Around Alone Race were nearing me. The head of the fleet was Marc Thiercelin on *Somewhere*, which was in line with me but much further south on its way to rounding the Horn. Following closely behind him was Isabelle Autissier, with Giovanni Soldini on his boat *FILA* behind her.

About ten days from the Horn I received an email from Mum telling me Isabelle Autissier had let off her 406 MHz EPIRB. My first thought was of the kind of weather she was having. There didn't look like there was anything around that could be a problem other than the usual 30-knot fronts that come through that area very quickly.

That night on the radio sked with Neil and the others I asked what was happening. No-one knew, not even the race organisers. The rescue control centre, wherever that was, had picked up a distress signal from Autissier's EPIRB, so they knew her position but there was no way of finding out what was wrong until someone actually got there.

Those boats were sailing in probably the most remote area anyone could be. It would take days for a navy vessel to get to her so the only other option was for one of her fellow competitors to head to her aid. I knew Giovanni Soldini was probably closest to her and that he'd detour to check out what was wrong with her.

It made me laugh during some of the media interviews I was doing when I was asked how I was involved in the rescue operation and what I thought when I heard the EPIRB go off.

On to Everest: New Zealand to Cape Horn

For starters you can't hear EPIRBs, and secondly, I was 1000 miles away from Isabelle Autissier, and it would have taken me a week to get to her. That's the media for you!

The hours counted down to Soldini's estimated time of arrival at Autissier's boat. I woke the next morning to an email from Mum saying that Isabelle's boat had turned upside down after her keel was snapped off by a wave and that Soldini had rescued her after throwing a hammer on her upturned hull to let her know he was there. They then left her boat floating in the ocean still upside-down and headed for Punta, in Chile, the next scheduled stop in the race, where she could be set ashore. Peter and I joked about all the equipment on the boat and Peter claimed he'd turn around and head back to the boat to grab the winches!

Autissier's keel snapped off after her electric autopilot failed and caused the boat to turn side-on to a wave. It broke onto the boat and caught the keel at a bad angle, snapping it off. The weight of the mast flipped the boat and it stayed that way until Soldini came to the rescue.

Speed was the only criteria on which racing boats were designed. They were incredibly light with as little drag under the water as possible to give them a plaining hull. Their keels were built from fancy materials which they hoped would have an edge on the other boats. It's no surprise that some break when pushed too hard trying to get to the finish line first.

Lionheart, on the other hand, had a much stronger keel, more integral to the hull. She was very much slower than the racing boats, but I wasn't there to set a speed record. All I had to do was take it easy, not break anything or fall overboard. There was no way I was going to push my boat. Mine was a marathon journey, not one of speed, and I was very aware of it.

At the time of the rescue I kept an ear out for any radio

traffic across the air waves in case there was an update. I managed to pick up two other competitors who were in between the leaders and the 40 footers a couple of weeks behind. They were Brad Van Liew on board *Balance Bar* and Mike Garside on *Magellan Alpha*. They were ahead of me closer to the Horn. I spoke to them a couple of times but usually I couldn't get through.

> *Tuesday, February 23*
> Roger says I can step on it now and go as fast as I can. Dropped storm jib as wind was 15–20 most of the day and unfurled genoa and raised double-reefed main. Heading due south at 6–6 1/2 knots. This should get me to 45°30'S by the morning which is where Peter picked up all his current. The Around Alone Race weather people are predicting some terrible weather next Tuesday and Wednesday when I expect to be at the Horn, but Roger says it doesn't look too bad just yet.
>
> *Wednesday, February 24, 12.45 p.m.*
> Peter rounded the Horn at midnight Melbourne time last night and I expect I'll round hopefully in 4–5 days' time. He said it is like there is a line that you cross when you get around the Horn— the swell dies right down to about a metre and it almost feels tropical. Sounds good to me.

Peter had rounded the Horn that day at 8 a.m. local time after slowing down during the night so he could pass it during daylight. I'd never contemplated that I could possibly pass the Horn in the dark. Imagine not seeing it! The weather was

On to Everest: New Zealand to Cape Horn

looking good so Roger gave me the go-ahead to get to the Horn as quickly as I could before the next low pressure system came through. The Around Alone Race headquarters sent me an email asking me about my safety equipment in case I was called on to help in a rescue.

> *Wednesday, February 24, 5 p.m.*
> The radio sked has just finished tonight.
> *Magellan Alpha* and *Balance Bar* moved their sked forward one hour which is the same time I talk to Neil and the others. Everyone ended up talking, including Peter who woke himself up especially and made himself a cup of tea.

The next day I had an amazing feeling when listening to the BBC. On the program they were discussing the modern 24-hour society and were interviewing the manager in the Coles supermarket at Sandringham. It totally blew me away. Here I was, listening to a world-wide station that interviewed people like Bill Clinton and leaders from nearly every country in the world and they were talking to the manager of a supermarket I used to walk past on my way to the Yacht Club. How about that!

The wind started to build to a force eight gale and I wasn't liking it much. It was coming from the northeast and I was only able to make headway south. It was taking me further away from where I wanted to go so I hove to, pointing the boat 60 degrees into the wind with a triple-reef mainsail up. I then jumped into the cabin, and waited for the bad weather to pass.

> *Friday, February 26, 8.30 a.m.*
> Wave nearly knocked me flat last night. Not a proper knockdown though. Wind was coming

down to 25 knots every now and then, but has picked up again.

Spoke to mates which was great. Made me all excited for some reason and I couldn't go to sleep. Going to bed now as it gets light.

A short time after the morning skies started to get light, *Lionheart* and I experienced our first proper knockdown which must have taken the mast past the horizontal point and into the water.

Friday, February 26
Just been knocked down a beauty. Stuff everywhere and also wet. Pain in the arse. Get me out of here!

I woke up just as the boat was coming back up, and all I could hear was water gushing down from the front hatch and the companionway slide. I reckon I must have gone past 90 degrees, as the microphone of the HF radio had come off its vertical slide holder and the navigation table had come open. It looked like a bomb had gone off in the cabin. The navigation books had been flung from the navigation table, all the food bags were thrown to starboard, all the ropes were out of their bags, a DD-size battery was lodged in the netting above my bunk, the flooring had dislodged, the leeward spinnaker pole was undone and banging about, the horn cleat tying the tiller off had broken, and some flying Tupperware had cracked the teak cupboards beside my bunk.

It happened so quickly there wasn't much time to be scared. The boat had returned to its usual position, lurching as it was buffeted by the conditions. It was as if nothing had happened,

On to Everest: New Zealand to Cape Horn

except that my normally cosy cabin was dripping wet, smelly and a total mess.

The rest of the day was terrible, but slowly the wind died down. That night I was able to raise the storm jib with the triple-reefed mainsail and started moving quite well. Later on, the wind died down to 15 knots so I raised the genoa and went to bed eating lollies and roll-ups.

During the nightly sked I spoke to Brad Van Liew who was 180 miles closer to the Horn. He'd had 50 knots gusting to 70 with 12-metre seas. He had been terribly worried about losing his mast as he was constantly getting knocked down.

The wind had dropped to a comfortable speed and the sun was shining the very next day. I really appreciated the chance to clean things up, except when I accidentally stepped on a carton of milk, splurting its contents across the cabin and all over my equipment. I scrubbed furiously at the mess and got into the mould that had been hanging about the corners for the last few months, singing along to Pearl Jam.

'Ooohh Iiiiiiiii OrrrrOrrr I'm still a-live!'

You can do that when you are at the end of the earth and no-one can hear you. It was funny, but the times I felt the most alive were after I'd pulled through bad weather.

The temperature was getting down to 6–7 degrees at night and I was only four days away from Cape Horn. The wind was light but still enough to keep me moving. I hadn't been able to make contact with Sydney Radio for the previous three days so I couldn't make any radio calls to home. I was still in contact via email and could always use the satellite phone for voice if I needed to, but I preferred to keep that for times I really had to.

Sunday, February 28
Position 55°04'S, 77°07'W.

> Not far to go now. Seas are great. Hope they stay like this! Couldn't get through to Sydney Radio so I tried Cape Town and Portishead but had no luck with either. I'm going to bed now as it is already getting light. I must convert my active hours so I'm up when it is light. Goodnight.

Since I left I'd been keeping to Melbourne time even though the part of the world I was in was actually fifteen hours behind. Cape Horn was actually on the same time zone as New York. I'd sleep late in the day and stay up at night so I could receive my family and friends' email and radio calls as soon as they placed them. I was willing to throw my schedule out of whack to communicate, as that was my lifeline. It also made media interviews easier, rather than waking up in the middle of the night and immediately talking to a radio station with thousands of listeners. I had enough trouble sounding coherent at the best of times, without blurting out something stupid because I was half asleep.

By the time I got close to Cape Horn I was going to sleep at 9 p.m. Melbourne time when in fact it was just starting to get light where I was. Being awake at night also meant I couldn't see the bad weather or the overcast skies. Out of sight, out of mind.

> *Monday, March 1*
> Today was fantastic. Wind has been varying from 10–20 knots and has swung around to the west–northwest. Got up a couple of hours before the sun went down but it was a great sunset, the first in a long while, and no swell means the cockpit is dry.

On to Everest: New Zealand to Cape Horn

That night was the second last before I got to Cape Horn.

> *Monday, March 1*
> Was up all night and didn't get much sleep as boat kept luffing up (pointing too close to the wind), so I was constantly fixing her course.

The swell started to die down as I closed in on land. I could sense land as I got closer, although I still couldn't see it. I'd been up during the night but was forcing myself to stay awake all day. I had to constantly monitor my progress and the changes in the wind so I could position myself to get as close to the Horn as possible.

> *Tuesday, March 2*
> What a start to the day. Didn't get much sleep this morning and noticed dolphins as the sun came up over the first land I'd seen since New Zealand 60 days ago.
> Man, it looked great. Was all hazy and mystical then this rainbow came out. I am overwhelmed with the beauty around me. One of the most beautiful days of my life. Only 13 hours approximate to the Horn. Yippeeee!

The way I felt was incredible. It was probably due to a lack of sleep and the excitement of the countdown to Cape Horn, but everything just seemed so amazingly beautiful. I sat in the cockpit with an iced coffee and felt the cold on my face as the sun rose above the horizon. The dolphins swimming alongside seemed to sense my excitement, escorting *Lionheart* and me past the first land—Islas Ildefonso. The cold air seemed to magnify the sights around me. I kept the boat speed to a maximum, considering the lightish winds, trying to get to the Horn before sunset.

But as the day wore on I realised it wasn't going to be possible. A dense cloud cover came in later in the afternoon and I decided to slow *Lionheart* down to arrive at the Horn at sunrise the next morning.

Apart from a few hours sleep and an interview with the Today show, I pretty much just monitored my speed and direction all night, plotting my position on the chart. About three hours before the sun was due to come up the wind died altogether, then swung to the northeast. I became a touch concerned when, for a while, I began drifting helplessly towards land on my port side. Thankfully, the wind kept swinging and went to the west, giving me some much needed propulsion.

> *Wednesday, March 3*
> Have been kept busy with sail changes. Am very tired now. Not much sleep in the last two days.
> Was visited by some type of animal at the back of the boat. Could have been a seal.

As the sun came up on my second day in sight of South America, so did the wind strength. It was down to 5 knots but then swung very quickly back to the east and picked up to a comfortable 10 knots. This meant I had to tack back and forth and the slow progress through the night meant I wasn't going to see the Horn until later that day.

> *Wednesday, March 3, 9.05 a.m.*
> Sun is starting to come up. I am seventeen miles away from the Horn. Picked it up on radar.

On to Everest: New Zealand to Cape Horn

Very soon after, I spotted its unmistakable shape. I'd dreamt about this moment for so long. The rising sun, combined with the early morning mist, made the sky a powerful orange colour, with the furthest headland only a small blob that came into view every now and then depending on the swell.

So this was Cape Horn. I could not imagine a better way to meet it! The Everest of sailing lay only a short distance ahead of me. I stood thinking of the aura of history and legend that surrounded this great cape, the Cape of Storms. I felt as though I'd been transported into some fiction book that Tolkien would write.

It was beautiful beyond description, and so real and clear to me that it felt in a sense, unreal.

To actually be there, at the Horn, created a sort of turmoil in my mind, as though I was grappling with two opposites. On the one hand there I was, a young kid who set out to do something many thought I shouldn't have, without knowing how either the boat or I would cope, and a résumé that said I hadn't finished school yet.

On the other hand was this legendary rock that lived, in my mind, on the same level as Ben Harper and Bill Clinton. The level that I only saw on television, that, as far as I could work out, was way beyond my reach, in another kind of dimension.

Yet, there I was. I had to quickly readjust my concept of where I stood in the world. I was learning the most valuable lesson of the trip: I could get anywhere I wanted to, no matter how impossible it seemed.

I was at Cape Horn, *the* Cape Horn! *The* Cape Horn that sailors throughout history had been terrified of and the subject of talk around Yacht Club bars. Hell, I may as well be jumping off the 5-metre diving board at Ringwood pool head first, or jamming with Ben Harper. Or maybe sharing a beer with the

President of the United States, or taking the first steps on the moon.

Hell, if I could sail my wimpy arse from the safety of the local Yacht Club, where everything was so straightforward and set out for me, to the kind of place on earth that you only read about, then I could do anything in the whole wide world!

YES, YES, YES! WHOO HOOOOOO!

I thank God for the incredible gift of confidence I'd gained. It was hard not be inspired when you'd just had the experience I had.

I tacked along the coastline comfortably, passing the snow-capped mountains on the land and lumps of thick kelp. The cloud cover had disappeared and gave way to the bright early morning sun that lit up the waters around *Lionheart*. The water was a different colour as it was much shallower. The wind swung around a little further to the northeast, which allowed me to take a better direction while I just sat and digested the thoughts racing through my mind.

> *Wednesday, March 3, 7.40 p.m.*
> Rounded at 18.00 zulu (universal time). Am totally buggered now. I haven't had a good sleep in a few days. It has all of a sudden just hit me— I've done it, and all by myself. I feel pretty good about rounding. I just keep staring at it, wondering how I could capture what it was like to show others. Feel a bit sad. Same as end of weekend feeling. Maybe it is because it signifies the end of one leg. Maybe though, it is loneliness because I am reminded about land and know there are people close by. It is easier when you are thousands of miles from anything. Couldn't call

On to Everest: New Zealand to Cape Horn

Mum and Andrew or Dad cos of poor coverage on phone. I'll be right though. Probably cos I'm tired too. Am feeling weaker and weaker every minute now. I think I just need a good sleep to settle emotions that are running high. A good sleep and I'll get with enough energy to tackle the next leg.

CHAPTER 7

Through a Mind Field: Cape Horn to the Azores

> Total focus. You, the sea, the boat, the sky are one entity. Expect hardship and discomfort but no time for boredom. Maintenance, study and writing; just do, do, do!
>
> — Note from John Hill

A definite pattern was emerging to my highs and lows. I was in the dumps again. It was the same feeling of emptiness that I had after seeing land and the fishermen off New Zealand. Yet, I felt the best after I passed through the hard times.

I suppose I should have seen this mental low coming. The Cape Horn build-up had had me on a high for weeks, since I started heading south on my descent for the rounding. It had been so large in my mind, not only because of its fearsome reputation, but because it was the place David Dicks had hit trouble and required assistance. In my mind, if I passed that point, I had a huge chance of completing the trip unassisted.

The three days before I rounded the Cape had been so amazing. There was this incredible feeling of anticipation,

Through a Mind Field: Cape Horn to the Azores

which, after I had rounded the Horn, had left me with nothing to replace it with. I was definitely experiencing what psychologists call the let down. I'm not sure if speaking to a psychologist before I left would have helped with this. I'm sure my mood was also intensified by a lack of sleep in the previous few days. I'd been running on excitement and adrenalin, but as that mix of fuel drained away, I felt lost.

I was writing my feelings down in my diary, and as I did I felt myself get weaker and weaker. The whole thing had been such a mental drain that when it was all over I just fell in a heap with the radar on, hoping the wind would stay constant and that there were no ships hanging about the area. I needed time to take in all the lessons I was learning and the strength to process all my thoughts and feelings.

I'd completed two of the five stages of the trip—Australia to New Zealand, and New Zealand to Cape Horn. I was on to the next stage—Cape Horn to my midway point at the Azores.

From Cape Horn onwards I assumed it was going to be all smooth sailing, but was I wrong! The weather improved dramatically as *Lionheart* and I headed up past the Falkland Islands. Then a series of fronts came from the west, giving me my second proper knockdown, and pushed me off course for about a week.

I was back on track mentally by the Falkland Islands, as I had something to focus my mind on. The *Herald Sun* had been trying to get a photograph of me ever since I neared Cape Horn. They'd contacted tour operators who travelled between Chile and Antarctica to see if anyone could take the picture. One operator was in the area as I rounded the Horn, but was heading south to Antarctica. If he did get a picture of me, he wouldn't be able to send it back to Melbourne for more than a month.

Lionheart

The paper's best hope lay with the Falkland Islands, and the air traffic to and from the islands. After many calls, the Royal Air Force base on the islands took up the case. But officialdom being what it was, by the time they got organised to do the flyover, I'd whistled past the islands and was out of range.

'It is ironic that a single-handed yacht sailing at 6 knots has apparently run away from our 300 mile an hour aircraft,' wrote Squadron Leader Gordon Parry, the Falkland Islands British Forces Commander in an email to the *Herald Sun* when they realised what had happened.

It was such a shame, as I was looking forward to seeing a huge military airplane do a low sweep over.

I lost contact with Peter while I was near the Falklands. He was well ahead of me, only about 20 degrees south of the equator.

> Monday, March 15
> I've got beautiful sunshine at the moment and yesterday I got so hot in the sun I had to strip down to thermals. My position is 44°11'S, 50°10'W.

With the better weather I could get into the jobs I'd been wanting to do for a while. There was a bit of movement in the wind vane which I couldn't work out. I got the tool kit out and realised that quite a few of the tools were going rusty. After I tweaked the nuts and bolts on the wind vane and slowed the movement down, I cleaned the tools and gave them an oil.

One job turned into another, which turned into another. During the knockdowns the cartons of milk under the galley bench had ruptured and were well and truly off. They'd started to affect the other cartons which had not yet broken. During one of the quiet days when there were a few rain clouds hanging about, I cleaned out the entire cupboard and separated the good ones from the bad. The milk had started to go hard and in some cases had actually set. I held the cartons that were broken or bloating over the side and punctured them with a knife. I then got all the empty and smelly cartons and threaded them onto a rope which I trailed behind the boat to clean all the gunk off before I stored them as rubbish.

I also collected all the dirty clothes I'd been wearing for the past three-and-a-half months and spent one sunny day washing them before rigging a clothesline that zigzagged all over the cockpit where I hung the clothes to dry. I then packed them away in dry bags. All the while I was sailing steadily north, up the coast of Argentina. What a pity I couldn't see the girls on the beaches!

Another day when there was no wind and the sun was beating down at full strength I got my wet weather gear out on deck to dry, then packed them away into garbage bags and stored them up the front with all the rubbish. I was to later discover the consequences of my actions.

There were many days of little wind, when *Lionheart* would sit still for hours on end. It was on one of those days, when *Lionheart* wasn't moving, that I spotted a pod of pilot whales. I grabbed the winch handle and started tapping it on the winch. A few of them diverted from their course and headed for me. I got the camera and watched through the viewfinder as one dived down and passed right under my rudder. Wow! I never actually stopped to think that one of them could damage the boat.

My days were a lot more comfortable by then. I was well past the Falklands and in radio contact with Neil and the others when they rounded the Horn. Neal Petersen on *No Barriers* saw his wind speed instrument get up to 77 knots in a storm passing Cape Horn. I was so lucky with my rounding, as I later found out that Roger Badham considered the weather surrounding me some of the worst he'd seen around Cape Horn:

> He [meaning me] rounded Cape Horn in some of the worst weather I've experienced in routing or following a boat race in that region. My solution was to keep him north of it, then quickly right into the middle, so as to minimise the nasty winds around some very intense lows. That was probably my best contribution to the voyage.

I was getting around in shorts and my tan was developing nicely. My water tanks had stopped working properly and were spitting out half salty water that I found hard to drink. I had 250 litres in jerry cans which I used for drinking and kept the yukky tank water for cooking.

Sunday, March 28
Am using second jerry can of water today.
Beautiful day. Feel a bit sick cos I sniffed some off milk. Is very hot day. Don't wear any clothes at all!

The next day the wind was fresh and *Lionheart* was moving quite well at 7 knots. The wind was just forward of the beam on the starboard side and the hot sun evaporated the saltwater, which was splashing up over the cockpit every now and then when a wave broke against me. I was keeping dry down below and had just turned on the BBC World service as I lay on my bunk. Not even a minute afterwards I heard and felt a loud thud on the hull somewhere forward of the mast. I assumed it was an extra large wave but I then heard the sound of the autopilot ram moving back and forth, which it did when it came off its attachment to the tiller. I immediately jumped up the companionway stairs and leaned out to reattach the tiller pilot.

I put the ram back in place then noticed a huge whale following behind the boat. I suddenly clicked as to what had happened and the adrenalin began pumping through my body. I'd read many stories of yachts that had been attacked by ramming whales and sunk for no apparent reason. At least this whale had a reason to be angry if he wanted to—I'd obviously run into him.

I didn't know what was going to happen as he was not alone. A second whale followed close behind. I had to get it all on film. If I was going to sink then there was nothing I could do about it, but at least I should try to film it, for the record. I whipped down below and switched the radio off. I grabbed the video camera, and turned it on while bounding back into the cockpit. The whale was still following, but much further behind. The camera was heavy and I felt that it wouldn't have picked up the whale in the water so I didn't bother. What a blow.

I stayed outside for a while longer to make sure the whales weren't planning a surprise attack from below, but nothing happened. I then checked the bilge for any excess water which would have indicated a leak somewhere. It seemed dry enough and *Lionheart* appeared to be sailing along as if nothing had happened. We'd both made it through unscathed.

I was stoked by the experience. Another possible disaster had been overcome. Bring on the next one, baby!

The strange thing was that I had a dream a few weeks before about hitting a whale. I started to pray that I didn't dream about colliding with a tanker!

Lionheart

The marine life seemed to be a lot more active in this part of the world. I soon found my first flying fish lying on deck. I was familiar with these strange creatures from the trip with Dave and there I was, on my own, with the small flying fish. It marked the beginning of the tropics for me. From then on they got bigger and bigger and the schools would sense the presence of *Lionheart* and take to the air in great numbers. At night I'd hear a thud followed by a few flickers to indicate one of the fish had landed on board and was stranded.

They were cute little fish but I soon came to hate the noise of one landing on board, for they played heavily on my conscience. I'd be woken by the thud of one ramming head-first into the boat then I'd try to get back to sleep. It just wasn't possible. The noise of them flipping about the deck screamed out to me to get up and throw them back overboard. It would have been easy if they died quickly but they flip-flopped away until the flips became less and less frequent, like the last gasps of a dying person. I'd fall asleep for a minute to be woken by another flicker of life. I couldn't bear lying there while I knew a fish was suffering. I usually couldn't get back to sleep until I'd got out of bed and thrown it overboard. It got pretty annoying when it happened three or four times a night.

Not only could I hear them land, but I could also smell them. The fresh smell of broken scales wafted down into the cabin where I could tell there was one on board even if it made no noise.

They could be a real problem. One morning I came out on deck and counted fifteen dead fish lying about the place: in the reefs of the sails, jammed beside the life raft and the wall, and even mixed in with the ropes in the rope bags. Some were quite large, up to 30 centimetres long. I tried to cook one once but there were too many bones and scales. At least I knew it was possible to eat them in an emergency.

As I headed further and further north it got hotter and hotter. I had to position my little 12-volt fan above my head and have it on all night just so I could get to sleep.

> Wednesday, March 31
> Roger has been predicting no wind for a few days but have had plenty. Then last night it died off and now there is none. Got up during the night and dropped sails cos they were annoying me so much. Sea is very flat and cos there is no wind it's very hot.

For the next three days I was becalmed. It was a taste of what was to come when I hit the doldrums. This was the area, 15° each side of the equator, notorious for days of no wind and stifling heat. It was said to drive sailors mad through frustration.

The days had already started to become very hot although I was still about 1000 miles from the doldrums. I was now about 500 miles off the coast of Brazil. I found the only way to get things done was to get up early and do them before the sun became unbearable.

> Thursday, April 1
> Don't feel the best, like I am wasting time (which I am) seeing that I am already behind schedule.

As the boat drifted around in circles so I too drifted around the boat, trying to escape the ever-present sun glaring down on me. I rigged up a shade out of my bunk sheet to read under but I couldn't enclose every side, so the sun still got in as the boat slowly drifted all over the place. I used the solar shower, which heated the water in no time when I hung it from the mast, to have a real good clean and a shave. I then put on some deodorant—all ready for Saturday night.

> *Friday, April 2 (Good Friday)*
> Still no wind! Saw a turtle. Changed lure and finally caught first fish—a baby dorado. Plenty for one meal but I used too much oil so it wasn't so nice. Also tried to make leather out of its skin. Rubbed salt and moisturiser into it when it started drying out and going hard. Such a pain in the arse and waste of time. Going nowhere. Water is oily calm.

The frustration was immense. Whenever the boat was sailing, everything felt worthwhile. I was making progress towards home, in a positive, forward direction. But when there was no wind and no movement, everything seemed grey. There was no point to being out there, no point to my existence. Time may as well have not existed. My actions had no meaning behind them because they weren't adding to the ultimate goal of getting home. Why was I feeding myself and wasting food? What was the point of getting up in the morning? Why do anything? There was no point to it. It was just idle time waiting around for mother nature to change her mood. I hated it! A slight breeze eventually picked up and I got moving again. It was generally light for the next five days, except for a huge rain cloud that passed overhead and sent the wind to 30 knots for about half an hour. It also dumped some of its load on me. I managed to block the cockpit drains and caught a whole heap of fresh water which I used to wash and wipe things down.

Through a Mind Field: Cape Horn to the Azores

Thursday, April 8
Have been buzzing tonight. First of all 3AW couldn't get through on the satellite phone so I called Megan (from the PR company) and she gave me the number to call them on. Did the interview, then called Mum at work to ask her to drop the tapes off at Trav's for his party tonight. Then I called Trav who said he just heard me on 3AW. Don't ask me why he was listening to 3AW! Ha! Goodnight.

That night I spoke to my friends at the party for more than an hour. Talking to so many of them really lifted my spirits. Anna and Katie told me how awesome Ben Harper had been at the Offshore Music Festival. I was jumping out of my skin with excitement. I wanted to be there among my friends, to have their vibes wash over me.

I estimated the call cost $900, which I felt really guilty about for days. I asked Mum to look into getting some kind of sponsorship deal with a communications company for cheaper calls. I hoped someone would come to the party because there was no way Mistral could afford for me to speak like that again.

Saturday, April 10
Wind is back—blessed are the trade winds! Read, ate, played guitar—that was my day basically.
The wind means that batteries are being brought back to full by the wind generator.

Sunday, April 11
Only new thing that happened today was while I was writing the column for the paper, I noticed

the fishing line was tight. I pulled in a beautiful tuna but it was half eaten by sharks. Don't know how long it had been there.

Monday, April 12
The tan is going very well—all over in fact—but there is a price you have to pay, moving about the boat extremely careful—you don't want to get anything caught in the rigging.

It was a much easier lifestyle in the tropics. The sailing was easier, if not a bit more frustrating, but at least I didn't need to be too worried about the onset of dangerous weather. The main concern in those waters was the tankers that pass between Europe and the Americas.

Tuesday, April 13
Spotted a couple of lights off port quarter.
Couldn't pick anything up on radar but I still think it must have been a ship . . . unless it's a UFO.

Those lights turned out to be a ship all right, and he was headed straight for me. I turned on the navigation lights and radioed him.
 Nothing. I tried again. This time I got a response, 'What you want?'
 He didn't sound particularly friendly. I tried to explain I just wanted him to know I was there but he didn't understand. He understood 'Goodbye, goodbye' though!
 I realised that as I entered the northern hemisphere I'd start to see more ships. I had to be careful as the large tankers

Through a Mind Field: Cape Horn to the Azores

couldn't pull up or quickly divert for something as small as *Lionheart*, which they probably wouldn't see in the first place. The maritime rule of power giving way to sail was not exactly the most practical rule out there. My worry was that I'd get caught in the path of a ship and have no wind to escape. It was a real danger. I was also concerned that the radar had not picked up the ship.

> *Sunday, April 18, 6.11 a.m.*
> Have just crossed into the northern hemisphere but I forgot to check if the water goes down the drain the opposite way. Will check next time I pass. ETA for the Azores is about 3–4 weeks.
>
> *Monday, April 19*
> I have lost all wind. Since crossing the equator it has been off and on. It has been bloody annoying as I raise and drop sails all day and then all of a sudden it's a squall of 25 knots from the opposite direction. Can't it give me a break? Swear at sails when they backwind. Punch them and have broken a few things. I'm rolling like a bloody rocking horse. Can't get away from anything. The whole rig makes such a noise when no sails are up and the rolling is such a pain in the arse. No sun, no wind, no power! (No fun)

I'd hit the doldrums. The frustration of not moving, combined with the dull, overcast and sticky, humid conditions, boiled up in me to the extent that I had to get it out somehow. I'd scream as loud as I could until I hurt my throat and collapse in tears and exhaustion.

I became very short-tempered, with many things taking the brunt of my anger. I took out my frustration on winch handles, the boom, torches and my water container, which broke and made collecting water from the jerry cans each day all the more difficult. It may sound like the tantrums of a spoilt brat, but it was an indescribable feeling. Relying on the wind, I was completely helpless when it failed to blow. Earlier in the trip when the wind failed to blow I felt I was wasting my time. I now felt as though I was trapped. I wasn't just wasting my time, I was being held hostage by the wind.

It stayed like that for five days, with annoying wind bursts, enough only to tease and frustrate me even more. Then, as though those previous five days had not existed, I hit the northeasterly trade winds. I was through the doldrums and had constant wind for the next couple of weeks. Whooo hooo! I was back in the good stuff and loving it.

Over the next few weeks I had my best run of the trip. I covered distances of 130 to 135 miles a day consistently. The wind was always on the beam or a little bit forward of the beam, causing the windward side of the boat to get a brown tinge. Peter had warned this would happen, though it wasn't a problem other than the sails getting dirty. They were very fine particles of sand being blown thousands of miles across the North Atlantic from the Sahara Desert. I could run my finger down a stay and it would leave a brown mark on the skin. When it rained, the dirt would collect in the puddles on the deck, forming its own little desert when it dried.

Saturday, May 1
I woke in the middle of a very vivid dream about the day I return home. I can remember looking at people in the street and having a

Through a Mind Field: Cape Horn to the Azores

very definite feeling of their presence (it sticks out in my mind cos it's something I miss out here).

My family was there. Everyone—Pop, Gran, Mandy, Stewart—they all made me feel so safe to be around them.

My batteries were getting quite low since two of the three solar panels had stopped working since the knockdown around Cape Horn. I couldn't afford to run the radar on watch for ships and I didn't have the navigation lights on for the same reason. I had to get up at night at intervals to check for shipping and also my course.

During that night I opened my eyes and wondered how long I'd been asleep for (one-and-a half hours, as it turned out). I woke from the dream wondering if my family and friends were actually on board the boat with me. I swung out of my bunk, wiped the hair off my face and looked at the wind speed instrument and the compass to check my course. I then received the fright of my life: there, behind the boat, was a huge tanker cutting across my stern on a perpendicular course, with all its lights on show. It was only a few hundred metres away and lit up like a Christmas tree. The first ship I saw in two-and-a-half weeks and it nearly hit me! I flicked on all the lights I had on board but was too afraid to turn on the VHF radio in case the captain abused me for not running my lights.

I had the feeling that because it was a full moon he'd seen my white sails and steered clear. That feeling was backed by the fact that when he passed, some of his lights went out. The tanker continued on its way and very soon was nothing but a single light on the horizon.

The experience was just too close for comfort. I resolved to leave my navigation lights on at all times, even if it did run down my power.

Sunday, May 2
Am annoyed cos Roger says that I'll be running out of wind for about a week. Has started now.

This wasn't good news. I'd just found out that Mum, Beau, Andrew, Gran and Megan from the PR company were going to fly over and see me at the Azores when I rounded the halfway mark. I was due there in a couple of weeks but the lack of wind was getting in my way and I disliked it with a vengeance. Still, there were little things that kept me occupied.

Monday, May 3, 10 p.m.
I have just seen the most amazing phosphorescence ever. I dipped the bucket overboard and the whole bucket glows. When I throw the water into the dead calm ocean it's like scattering a bucket full of fluorescent coals over the ground. The patch where this happened only lasted a couple of minutes even at the slow rate I was going (0.7 knots).

Tuesday, May 4
Once again, no wind today. Just drifted. Spoke to Dad last night on the phone, which made me feel great. Saw the lights of two ships in the distance last night. Another two today as well. Sun has just gone down and I am having brief visits by the phosphorescence again. It is strange cos one

Through a Mind Field: Cape Horn to the Azores

second it's there when I throw water on the surface and when I throw the next bucket in the same area (cos I am not moving) it's disappeared. It is a clear sky and the water is oily calm. The reflecting stars look like lost jewels beneath the ocean and two very bright planets are reflecting and look like the eyes of a deep sea creature staring up at me. I have bad images in my mind of a huge tentacle coming out of the water and pulling me down to where it came from, like something out of a movie. It is dead quiet and not much to do. I'm not very tired. I hope some emails come tonight.

A few days later the wind arrived again. Usually the area I was in had consistent wind but apparently, as I approached the Azores, a high-pressure system that usually sat permanently over the Azores moved towards me and becalmed *Lionheart*. It screwed up Roger's plan to make best advantage of the weather.

Each night I'd report my position to an American gentleman called Fred. He was on the airwaves every day, listening out for emergencies and helping yachts cross the Atlantic by monitoring their progress. The organisers of the Around Alone Race had put me on to him, so we spoke every night. It was sometimes a pain because I had to get up in the middle of the night at the time we'd arranged, which was evening where Fred was. My alarm clock didn't work as it was full of water from the knockdowns. Sometimes I'd miss our sked or I'd wake up twenty minutes beforehand and have to wait for Fred to come on air. One night after I'd finished giving my website address to Fred over the radio, another voice came over calling for a Venture Beyond vessel. I presumed he was referring to me, as that was my web address. I corrected him on my vessel's name and started chatting.

The caller, John, was on a boat called *Dragonfly*, which was relatively close to me, maybe 400 miles away. He was coming back from the Caribbean to the United Kingdom with another boat, *Midnight Getaway*, sailed by his friend Andrew and his wife. They were keen to know my story and what I was doing out there on my own. They, too, were headed for the Azores for a couple of days' rest so we organised a sked among the three of us. I loved the contact, which I'd badly missed since I lost contact with Peter and Neil and the others who had pulled into port in South America. Those chats with John and Andrew made me feel like I was back in life's mainstream, rather than out in the middle of nowhere, miles away from civilisation. It made me feel safer knowing yachts were constantly crossing the stretch of water I was sailing in.

Friday, May 7
I am now about ten days away from the Azores—my halfway point—and about to stare at my family face to face for the first time since leaving. A lot of things have changed already. My hair is now shoulder length, which itches my face and my skin a healthy brown, but more than anything, we will be looking at each other in a different light, a shared knowledge of experience that will bring each of them closer to me—something tangible that will link us. I will lock eyes with Mum and be able to relate as two experienced at washing clothes. My view will shift to Andrew and I'll smile with a better understanding of the guitar. Then I'll notice my brother

Through a Mind Field: Cape Horn to the Azores

and we will acknowledge each other's presence and care, which is so hard for brothers to do and it will all be because this trip has changed ME. I know how to handle a 34-foot pointed eggshell while veering down a wave half as high as the mast. I know how to be responsible for my own safety in the challenges this earth contains and I know that I can achieve anything I put my mind to.

But when I look into my grandma's eyes a week and a half from now, I'll be humbled by her aged stare and experienced face and the knowledge that there is still so much more to learn—after all, I've still got half a voyage ahead of me.

The excitement began to build as I neared the Azores. The weather was still favourable, even with the Azores high shifting towards me and becalming *Lionheart* for nearly a week. Mum and the rest were already there waiting for me.

Before I rounded the Azores I had to sail in an arc to pass through my antipodal point at latitude 38°18′N and longitude 35°22′W, the exact opposite point to my starting point at the Port Phillip Heads. A few days after that and I'd be at the Azores, where I would actually turn and head for home. I was certainly looking forward to that.

Mum and the others occupied their time sightseeing around Horta, the main port of the Azores on the island of Faial. One of their tasks was to organise a local official who would come out on a boat to hand over my mail under official conditions. He would also inspect it to make sure it contained only mail and nothing that could actually help me. They found a local solicitor to do the job. They'd also met up with a man, Altino, whom the *Herald Sun* had lined up to email Beau's pictures of me back to Melbourne to put in the paper. Altino had a good radio system, so I was able to talk to Mum as I got closer. We mainly just spoke about how far I had to go and how exciting it was. I asked what the islands were like and tried imagining them. Knowing there were cows and fields and people sitting in cafés less than a week's travel away left me with an indescribable feeling.

I continued my way northwards for the antipodal point and soon got higher than my fellow sailors John and Andrew, who were heading directly towards the Azores.

> *Saturday, May 15*
> Just caught a dorado—beautiful colours and good size. Had just finished speaking to Andrew after he and John caught a tuna each when I looked up at the line and I had one too. Have been thinking I will deep-fry him. Mmmmmm! Nearly at antipodal. Should be four days before I get near Faial. Have been seeing a few planes about, even during the day. I saw one today and there have been heaps of jet trails in the sky. Should be passing antipodal later tonight.

I'd caught my third fish, a one-metre-long dorado. It had the most beautiful rainbow colours through its scales. Andrew told me over the radio how to make fish cakes with fish and potato and other herbs and spices. He got me enthused to look through the galley cupboard for ingredients where I came across three packets of bread mix. Man, what a great feeling that was! There was nothing like the yeasty flavour of bread with canned jam on it.

Through a Mind Field: Cape Horn to the Azores

It was the day before I reached the antipodal point and I had good reasons to celebrate—sunshine and perfect conditions, fresh fish, bread, radio company and I was only a few days from seeing my family. It was as though all my Christmases had come at once, and I didn't know what I should stop and enjoy the most.

I cleaned the fish without the dedication I'd usually apply to the task. As I was going to make some bread afterwards there was no need to spread my activities out for as long as I could. I treated it as food preparation rather than a time-filler. I took a few short-cuts and didn't get all the flesh I could from the fillets. I was thinking of towing the carcass behind the boat to try to attract a shark or two but I was too busy and had other things to do so I just ditched it overboard.

The plan was to cook the fish for lunch, then make the bread in the afternoon and have it for dinner. I filled half the pot with oil and tried to deep-fry the fish. I made batter but it didn't work too well. The batter separated from the fish and it was a mess trying to get it out. It ended up not tasting too good, so I ditched what should have been a perfect meal. A man called Raphael, who lived in the Canary Islands and was monitoring my position for Mum and Altino, called me in the midst of my preparation, so I hurriedly took the call, covering the radio microphone with fish oil. I had so many activities on the go.

It got me thinking about how my new way of life differed from that at home. What made life at sea so special was my enjoyment of simple tasks. That day I was living my life like I was at home, rushing from task to task, not taking the time to enjoy or appreciate any of them. The clutter of activities in our everyday lives leaves us no time to enjoy simple things, like cooking a fish or talking on the phone. That day, with so many things on the go, I'd wasted the opportunity to enjoy each separate task because of the anticipation of what I would do next. It ended in disappointment.

> *Saturday, May 15*
> Position now is 38°12'N, 35°39'W.
> About 15 nautical miles away from antipodal point. South wind at 10–12 knots.
> 11 a.m. Melbourne time—38°18'49"N, 35°20'18"W. Just passed antipodal. Can now get some sleep after emailing everyone.

I stayed up late that night watching the GPS count down until I was north of the antipodal point. I then changed course to head east. I waited another hour until I passed it totally. I was happier that I could finally go to sleep than with the fact I was heading towards home. My reward was no longer the record. All I cared about was finishing what I'd set out to achieve and seeing the people I loved. I was desperately looking forward to our rendezvous.

The next day I was on a course about due east making a good 7 knots.

> *Sunday, May 16*
> Did record day's run today. About 165 miles.
> Wind south at 12–14 knots and I'm flying along. Cooked bread which is beautiful. Spoke to some Aussies on a delivery yacht called *Coconut*. Good speaking to them.

I cleaned up the boat in the morning and, in the afternoon, I got into the second bag of bread mix. I took my time and laid out everything properly. I warmed the water to the right temperature, kneaded the bread well and even waited an extra fifteen minutes longer than the instructed time for it to rise. I didn't have an oven so I made little flat roll thingies and grilled them with tender loving care under the grill, flipping them at regular intervals. I spoke to John and Andrew who had no wind and were motoring towards Horta. They expected to arrive there late that night. I also spoke to Mum who was at Altino's, but not before I'd made contact with Raphael to report my position.

I was so content eating my bread then headed off to sleep a happy boy, knowing I would soon see Mum and my family.

> Tuesday, May 18
> Man, it seems like it's taken forever since the equator. Wind is roughly southeast so I'm hard against it to get more south. Should be there tomorrow morning. Sailing is great—sunshine and good winds.

I was making my final bag of bread mix when I was visited by a pod of dolphins and a turtle with something growing on its back. I got as far as possible in the good winds then slowed *Lionheart* in the afternoon. I wasn't going to make it to my meeting that day but if I kept sailing at the pace I was going through the night, I'd sail straight past the small group of islands altogether. I put a few reefs in the sail and furled some of the genoa in until the boat was doing 4 knots. I enjoyed my bread like nothing else. I'd take a bite, strum some chords on the guitar, and sing with my mouth full. Another bite and another line of the song.

What a life! The wind eased up as it got dark, but there was still enough to keep me moving sufficiently. I spoke to Raphael and Mum and Andrew then went straight to sleep after signing off with the words, 'I'll see you tomorrow, love you, bye.'

I woke a couple of hours later to check my course and position. As I pulled my chest up through the companionway, I turned to the bow and was taken by surprise. On the horizon was a thin line of orange lights, signalling the coast of Faial.

I sat there looking at those strange shimmering things for a while, then checked the radar for any shipping before heading back to bed. I battled to get to sleep again, but I succeeded for a few hours. I woke again and rushed outside to find the lights were closer and spread further along the horizon.

The sky had taken on a blue tinge in the far east, meaning the sun would soon be up. There was no way I could get back to sleep so I got dressed and manually steered as the sun came up over the next hour or two and the sky put on a show of red clouds.

Soon the land came into view and immediately cast a wind shadow as I sailed closer to the island. The slack sails started doing their annoying thing of slapping all over the place. I persevered in steering the boat, trying to eke out every bit of distance I could. This went on for about three hours until the wind picked up and died again.

By mid-morning the gusts were getting stronger and stronger and the day had turned to a dirty grey. One minute there was a 5- to 6-knot breeze, a few minutes later it was up to 20 to 25 knots. What made the situation worse was that I had to make my way to the Horta marina, situated in the passage between the first island I'd already passed and the second which was right next to it. To travel up this passage I had to sail directly into a headwind. Add the choppy sea from currents around the islands and the gusting wind which kept changing direction and there was a good enough reason for me to be very angry with the weather.

I dropped and raised sails constantly, trying to inch my way in the right direction. But I made no progress. I called Mum on her mobile phone to tell her about the problems and that I couldn't get to where they were. By lunchtime I'd failed to arrive so they decided to leave the marina to find me. I gave them my position, then sat back with the binoculars, eagerly scanning the direction they should come from.

Time dragged on and the gusting continued, growing more frequent and lasting longer. I began drifting towards land, forcing me to pull some of the furler out to get away from the coast. I then heard a call for me over the VHF radio which I answered immediately. I discovered why they were taking so long. The two boats they were on didn't have charts, so they were guessing my position. I gave them my latest position, three miles off the northwest coast of Pico, and they told me to stay where I was. I was getting quite upset about the way things were going.

Things just weren't going the way I'd hoped and imagined they would. I'd expected the waters to be a deep blue colour with the sun out and a gentle breeze so I could stop *Lionheart* and speak with my family from sunrise to sunset. I imagined myself going down below to cook my own lunch while they brought out sandwiches which I'd stare at in envy. It just wasn't turning out like that. I made three more phone calls, giving my position over and over again. Mum and Andrew were fighting a language barrier as they battled to tell the captains of the boats where I was.

Sometime that afternoon—I can't recall what time, just after I'd explained my position once again—I heard someone yell, 'There he is,' over the phone as I spoke to Mum. A big white fishing boat, probably 15 to 20 metres long, was heading directly towards me. As it drew closer I could see my grandma, two cameramen, two reporters, the captain, Altino and Megan, who had organised the media from Australia. I couldn't stop smiling as I waved to Gran, Megan and Altino. There was no sign of Mum.

The questions started from the media. It was difficult to speak because of the distance between me and the boat, coupled with the weather conditions and the reporters' Portuguese accents. I definitely heard Megan ask me if I'd seen her spew over the side of the boat, while Gran just asked how I was, and smiled as she waved.

I then saw a dinghy approaching with several people in fluoro orange jackets sitting on the edge of the hulls. One of them was Mum. I swung my arm into the air and waved slowly. They did the same as they continued towards me. I just stood there watching the familiar faces, strangers in this environment. They got close enough to start talking, with dozens of questions flying across the water.

Except for Mum. She just held her hand over her mouth and had tears flowing from her scrunched-up eyes. There were six people crammed into the dinghy. Beau was up the front, being very business-like as he took photos alongside another photographer. Andrew was behind them, holding the video camera, the captain operated the controls while Mum stood at the back with the local official who would hand me my mail and make sure no-one touched the boat or handed anything else over. As we'd been in radio contact for more than a week and, of course, emailing nearly every day of the journey, there was no need to blurt out a million details about what had happened in the previous five months. It was simply a chance for me to see my family, and for them to look me in the eye to work out if everything was going as well as I said it was.

Through a Mind Field: Cape Horn to the Azores

I stood there and smiled, and probably gave them the impression things were as good as I'd told them in the emails and calls. Mum settled down after about ten minutes so we could talk normally. I'd like to be able to write what we talked about but I honestly can't remember and it's probably not that important. The main thing was that I got rid of my rubbish to them, I received my mail, and Mum and I got to look at each other. Just on 40 minutes after they arrived, they had to leave. The wind was getting stronger and they'd spent so much time trying to find me that they were running low on fuel. The fishing boat left first and soon after, so did the dinghy. Mum started crying again but I reassured her that everything was going great and that I was very confident in myself and the boat and that this second half would go much quicker than the first, after all, I was going downhill.

Mum still gets upset when she talks about that moment. She said the look on my face as they turned and left was heartbreaking for her. It was all she could do to not turn around and come back to get me. The dinghy pulled away and we kept waving. I went down below and put the package of mail on my bunk, then came back up and stood halfway out of the cabin, waving and watching them continue to bob over the waves until they disappeared. My family reunion had not turned out the way I'd hoped it would.

~

Again, I should have been ready for it, but I wasn't. An overwhelming feeling of emptiness hit me. The Sunday afternoon feeling had returned again, made worse by the afternoon sun putting on a beautiful display as it fell towards the horizon. My family was heading back to a hotel room to have a warm shower and something nice to eat, yet I couldn't be a part of it.

I'd been living in a wonderland in the build-up to seeing them. It had become another Cape Horn all over again. I looked at my surrounds, at the filthy salty squalor of my cabin. My cramped conditions would surely be deemed unsuitable for human inhabitation. I'd just finished telling my family how much I was enjoying it, yet at that moment all I wanted to do was be with them. I suppose it was just the natural law of nature—what goes up must come down—and I was coming down fast.

    I sat on my bunk, not caring which direction the boat was sailing as I opened the mail from friends and family. The letters took me back to what it was like at home. There were stories about school that seemed unreal, and nice things people had written. I was especially taken by the effort my usually raucous friends put into writing special letters with real meaning. I'll treasure those forever.

CHAPTER 8

Please, God, Stop this for Me: Azores to Cape of Good Hope

Thursday, May 20
Last night the wind was terrible around the islands and the current made it quite choppy. I tried to get through channel but couldn't so headed back the way I came. Read mail and hove to at night cos it was so shitty. All I can do now is keep moving and get home as quick as possible. Would help if there was wind though.

I found that after a couple of months in the tropics, where the wind didn't get much stronger than 15 knots, a 25-knot front seemed terrible. So terrible that I chose to heave to. In the Pacific or the south Atlantic, 25 knots would be the average wind strength. But as I'd been in calm weather for so long, a strong breeze took on amazing proportions. Admittedly, the currents around the islands messed up the waves so they knocked me about quite a bit, but still, the conditions were nothing I couldn't expect to have daily once I got back into the Southern Ocean.

The morning after I saw my family the wind was still 25 knots, but as the day wore on it died down, until lunchtime when it dropped under 5 knots. I wasn't moving at all.

> *Thursday, May 20*
> Waited all day as swell died down while not moving anywhere. Still no sail up tonight as there's no wind.

The next day the wind picked up a little, to 8 knots, just enough to keep me moving. At least with the flat ocean the sails didn't jar the rig. The sun came out after days of overcast conditions, so I took the opportunity to tighten the shrouds that held the mast in place. It was just another example of how the weather affected my mood. The previous few days of feeling down since leaving my family suddenly disappeared and gave way to a feeling of achievement as I tightened the rigging. But the feelings were not to last too long, as the clouds reappeared.

> *Friday, May 21*
> Wind just died again and cloud cover came over. Can still see Faial which is a major shitty start for my leg home! Saw four whales and one real close up. 2.8 knots of wind. Yuk. No sail up now!

I still had a small amount of sail up and was gliding ever so slowly over the grey water. My roller-coaster of emotions continued on its merry way as I again grew frustrated looking at the blob of land seemingly stuck behind me. Over to port I noticed a whale swimming alongside. I edged the tiller to slowly move closer until *Lionheart* was beside the whale. It must have sensed my presence, and dived into the depths, never to return.

Over the next half hour I spotted another two of them through the binoculars.

As the late afternoon sky was seeping a few of its colours through the grey clouds, I saw the fourth whale of the afternoon, with its back just out of the water. This whale was ahead on the starboard side, swimming on a perpendicular course that would bring our paths together so we could get a better look at each other. I stood on the bow of *Lionheart* holding onto the furler and looking out at the big creature closing in. A minute passed and we continued to get closer and closer. I realised there was a growing chance we'd collide.

My mind raced over what to do. Should I run back to the tiller and alter my course to pull away from the whale or should I enjoy every second of this once-in-a-lifetime chance to get close enough to a whale that I could jump on its back if I wanted? I was transfixed as I watched the mammal get closer and closer. What should I do, what should I do? I could see its full body through the clear water. It moved so slowly and peacefully that I decided that if we did collide, it would be gentle enough to prevent injury to the whale or *Lionheart*. Luckily I didn't have to worry about it. I leant out over the bow as the whale passed only a matter of metres in front of me.

He must have sensed my presence because suddenly he stuck his back out of the water and took a long breath then pulled his nose under and dived down into the depths, giving a small flick of his tail and leaving only the turbulence in the water and the moisture from his breath. The picture is still so clear in my mind—I'd made the right decision!

Lionheart

For the next three days the wind picked up then died again, which meant I was still able to see the pointed volcanic island of Pico. That rubbed salt into the wound even more. My family had been waiting at the Azores for nearly two weeks while I was becalmed, then the meeting was so rushed and not how I'd imagined it would be. And there I was, staring at the island while they were back home again. The timing just didn't seem right, which was the most frustrating part of it all. I could have spent those last few days speaking to them from another boat rather than just drifting aimlessly on my own wasting time. It was so hard to take.

Monday, May 24
Wind increased until it was 15–20 knots. Did a bit of tidying up and listened to Bob Dylan. Nice sunny day. Saw some more dolphins. Received emails from mates via Barbara late last night. Had a coffee in the afternoon so I didn't get to sleep until 2 a.m.

I soon made it into the northeast trade winds, which were perfect as they came from behind and were always consistent, with sunshine most of the time.

Friday, May 28
I'm surely in the trade winds now. Beautiful conditions and easy sailing. The nights are great too. It's warm enough to sit outside in the cockpit wearing nothing with all the lights turned off and look up at the stars, which seem to be everywhere. The positions of the constellations are so familiar that I can tell when I am

Please, God, Stop this for Me: Azores to Cape of Good Hope

off-course with a glance to the sky, depending on what time of the night it is.

I started to spend many nights outside as the weather got warmer and the boat was moving well without water spraying over the cockpit. All my torches had broken, either from running them all night to light up the sails in case of nearby shipping or from saltwater. I still had heaps of D-sized batteries left so I bent some wire and used gaffer tape and a globe to make my own light, which I strapped on to the solar panel frame. It would light up the whole cockpit with a beautiful warm light like that from a candle. I'd sit out there at night with my guitar and something to nibble on and dodge the occasional rain cloud by jumping down into the cabin. It was a great time of the trip. I read books like *Survive the Savage Sea* and *Fatu Hive: Back to Nature*, which kept my imagination working about someday stopping off at all the interesting places I read about in these books. It was a time of great dreaming.

Monday, May 31
Position is 27°38′N, 26°21′W.
I've discovered a new hobby—picking the sultanas out of my cereal cos I don't like them! It has everything a good hobby consists of—a challenge, a reward and best of all, it passes the time.

Thursday, June 3
Flying fish reappeared all of a sudden two days ago. Last night had about twelve land on boat so I was up putting them back in the water. Have been moving along great the last two days.

Should do approx. 150 miles today. Got good news from Mum this morning that the call costs from Iridium are dropping next month from $15 per minute to $4.45. Great! Worked out on the big chart that I need to average 5.5 knots (135 miles a day) to be back in time for my birthday. Will try for this but I doubt it will happen. Most probably ETA is mid-September.

In a couple of days I would have been out here for six months. That's HALF A YEAR. Doesn't seem that long. I really hope I can get home in three months like originally planned. Don't want to be out here for another half a year.

It wasn't that I disliked it out there. I was having a great time and living a very satisfying life. But I was missing the normal things of home. If I spoke to someone on the phone for ten minutes that was a huge amount of time to be talking to another person, and it made the rest of the day great, even if the conditions weren't so perfect. Yet, if I was at home I could call a mate up for 30 cents and speak for as long as we wanted. I looked forward to human interaction when I got home. I'd estimated my trip would take roughly nine months to complete.

I hadn't worked out the distance and how fast I'd be going, instead I just took how long it had taken David Dicks—264 days—and expected that I'd do the same.

It looked like I'd make it home about mid-September which left me about three weeks up my sleeve before I got home older than David Dicks. I had to be home by 6 October to claim the age record. While out at sea, the record hadn't weighed on my mind all that much. David got back at age eighteen years and 41 days and I turned that on 6 October.

Please, God, Stop this for Me: Azores to Cape of Good Hope

But David hadn't completed his voyage unassisted. The youngest person after David to complete a non-stop trip unassisted was a 26-year-old, so I had plenty of time to get home in time to break a record.

The record meant nothing to me when I was out at sea. So what if I got home with or without a record! It was nothing in the scheme of things. I was looking forward to the hugs and kisses and the human contact far more than a piece of paper that I could use to show people something I already knew myself. I'd have very happily traded a can of Coke for the record.

You might ask why I didn't. Why didn't I ask Mum to hand me more supplies or stop off and see the sights? There were two reasons. First, I had commitments back home to sponsors who had put a lot into my trip, and secondly, because I wanted to finish something I'd started. This was probably the main reason. If I pulled into land or succumbed to the temptation of accepting assistance when there was no need for it, then I'd have chosen not to succeed. From the very beginning my aim was to finish, and that was my choice. Either I couldn't because of circumstances out of my control or because I gave up. It was as straightforward as that, and I was simply giving it my best shot.

If, for whatever reason, I couldn't finish because things were out of my control, then it didn't matter that much. As long as I did the best I could, I was satisfied and that's what the trip was about—being satisfied with myself!

> *Wednesday, June 9*
> A bit of adrenalin pumping around the old veins at the moment. It's been a slow, calm, hot, sunny day and I've been in the cockpit reading. I stood up just a little while ago to take a leak overboard when I heard an unfamiliar noise. I looked out

over the starboard quarter and there was this boat heading straight for me several miles away. It was only a fishing boat but I was concerned cos this area is pirate territory. How often does a boat in the middle of the ocean start heading straight for you?! I jumped down below and called out on the VHF while trying to get into some shorts at the same time. I got an answer, but they only spoke Spanish (I think) meaning it could still have been a pirate boat. I furled the genoa and changed course to see if they would follow but they were really close by now. Five crew stood on deck and started waving madly and taking photos . . . Phew!!! I waved back and whipped down below to get my camera while we both took photos of each other. Then they kept on going so I unfurled the genoa again.

I figured they must have either heard me reporting my position to Fred each night and knew what I was doing. Or else they were really homesick and were glad to see another boat, as we were hundreds of miles from the coast of Africa. I was also glad to see them after I realised they were no threat to me.

Not long after that, I had another shipping encounter. I was tracking due south, with enough wind to keep me happy, but it had a nasty habit of dying every now and then.

I clearly remember the wind dying during the middle of one night so in my sleepy state I dropped the mainsail about a metre to take the bite out of it whipping and shuddering the whole rig. The sun was coming up as I got up to check things, however I decided as I was not moving I may as well go back to bed and have a sleep-in.

My slumber was shattered by three long loud honks which

Please, God, Stop this for Me: Azores to Cape of Good Hope

scared the living daylights out of me. I cleared the lee sheet holding me into the bunk and scrambled naked up the stairs, cleaning the sleep from my eyes and trying to focus on where the noise had come from.

About 100 metres away was a huge tanker heading in the same direction as I was, parallel to me. Seeing that there was no immediate danger, my mind jumped to the next priority, which was to cover myself from the sailors who were no doubt watching. I swung down below and got into some shorts before coming back up again and waving in the general direction of the ship. I couldn't see anyone but I suspect they saw the limp sails from a distance and with no-one on deck on such a warm sunny morning, became concerned for my welfare and detoured to check it out. That's the law of the sea for you. It gave me a good feeling to know 200,000-tonne strangers were looking out for a small boat like *Lionheart*.

I was getting closer to the equator and Roger was trying to work out the best longitude for me to cross through the doldrums. He'd been keeping an eye on the weather patterns around the area to provide the best route to get me through the area as quickly as possible.

I expected to be stuck in the windless area for some time but hopefully the stories about yachts being stuck for weeks on end wouldn't happen to me. My first crossing, when heading up towards the Azores, had been pretty quick and I hoped this one would be similar. At least one consolation was that I knew I had less of the trip to go than I had already done, which would make any delay easier to handle.

The first signs of the doldrums soon appeared. As I neared the equator again, it became very humid and muggy, and the salt that penetrated all my clothing and bedding attracted moisture, making everything damp. Clothes were so uncomfortable, I went without most of the time, except when I was video-recording. Even then, there were a few times I forgot, catching myself on camera.

But the major problem was that the wind simply disappeared. Then came the huge rain clouds which dumped their loads with such intensity that the rain would envelop the boat like a thick fog. The wind blew up to 30 knots in those sporadic gales. As soon as the cloud passed, the gusting wind blew from all directions, changing constantly and chopping up the swell and rocking *Lionheart* all over the place. It made life very uncomfortable. It would be very grey and overcast during those rain clouds, yet half an hour later the sun would be out, beating down in full equatorial strength, highlighting the dark clouds surrounding me, with no wind at all.

The rain gave me a chance to replenish my fresh water supplies. I'd sit in the rain with a bucket in one hand and a sponge in the other. The rain would hit the rigging and sails and drip down to the deck, draining towards the cockpit where I sponged it into the bucket until it was full. I made a funnel from an empty milk carton, decanting the water from the bucket into a jerry can. I managed to fill one-and-a-half jerry cans in this way, which turned out to be a bit of a lifesaver by the end of the trip. Things would have been pretty desperate in those final few weeks without it.

> *Monday, June 14*
> Moving very slow. Went backwards today! My pos is 05°46'N, 25°12'W.

For five days I rocked and wallowed hopelessly in the doldrums. It was another waste of time, except for an unusual experience with a pod of dolphins.

Please, God, Stop this for Me: Azores to Cape of Good Hope

The sun was out and there were a few small rainbows hanging about the ever-present rain clouds. There was no swell except for a few small waves created by the last downpour. I was lying in my bunk when I heard a noise through the fibreglass hull. I assumed it was the clicks and whistles of a pod of dolphins approaching the boat. I went outside, checking the piece of wool tied to the shrouds for any hint of wind. It was limp, with not a breath in the air. I looked over the side of *Lionheart* and saw more than a dozen dolphins swimming around the boat. I was intrigued, as they behaved like a human family. Every other time I'd seen dolphins on the trip (which was quite often), *Lionheart* was sailing along, with only the biggest of the pod swimming in the bow wave, surfacing for a breath every now and then. But this time the pod had come over to look at me even though I wasn't moving. I started talking to them, which sounds stupid, but it seemed the most normal thing to do out there.

'Hey guys, do a flip for us, come on,' I'd say, or I'd try to mimic their whistles, with poor results.

I could distinguish the older ones with their scratched torsos and pieces of fins missing, no doubt as a result of close encounters with sharks or other big fish. The younger dolphins were much smaller, with smoother torsos. They'd obviously not had their scrapes and fights, which were sure to come. The younger ones swam as a unit just behind their mothers and copied every turn and movement their guardians made.

Then there were the aggressive ones, who I assume were the chiefs of the pod. They would leap out of the water sideways and whack the water with their tails and hind, making a big splash, then 10 metres later, do the same thing five or six times over.

But by far my favourites were those I called the teenagers. They would gracefully get a run up under the water then leap clear of the surface while spinning like a torpedo four or five times, belly-flopping down into the water with a splash. It was so spectacular, happening only a few metres from the boat, like something you'd see at Sea World.

They reminded me of my mates and me, going to the pool on hot days on our skateboards to cool down and checking out the scenery. We'd energetically attempt flips and strange moves off the diving boards, often hurting ourselves, all in the name of impressing chicks.

The dolphins stayed with me for several hours, while I sat fascinated by their antics. But as it started to get dark, they left me again, to battle the frustration of the doldrums alone.

I crossed the equator again on 21 June, but the milestone went unnoticed. I was going to check the water draining down the sink, and didn't even realise I was back into the southern hemisphere until well after the event.

The rain clouds slowly disappeared as I drifted, then sailed slowly south when the breeze became more frequent. I'd made it through the doldrums in relatively good time and hit the southeast trade winds.

I'd only used the mainsail on its own over the last week, but with the more consistent wind the genoa was able to come into play. It got about a third of the way out when it wouldn't move any more. I went up the front and tried to untwist it myself, but the furler just wouldn't budge. I was able to reverse the sail, pulling it back in, but the furler made a terrible grinding noise. I went below and grabbed the magic WD40 lubricant spray. I sprayed the furler a few times and it started to move again, but still with some friction. I presumed some dried salt had settled inside the casing as it sat dormant during the past

Please, God, Stop this for Me: Azores to Cape of Good Hope

week. I continued to work the mechanism until the drum began spinning more smoothly.

I went back to the cockpit and started to pull the sail out again until it pretty much got to the same point as before and wouldn't move any more. Then I did something really stupid, something I'd been trying not to do the entire trip. I used force, thinking that was needed to turn the dried salt to powder. I gradually winched the sail all the way out. I was moving quite well so I set the boat to steer itself then went up forward to give the furler a couple more sprays.

It wasn't until well into the afternoon, when I had to furl in some of the genoa as the wind picked up, that I noticed the drum had come apart from its base. It was pulling upwards, providing less tension for the sail. My first thoughts were that this was a disaster.

I quickly tested the furler and discovered I was able to furl the sail in although, without the tension, it began to kink as it was wound around itself. I had no choice but to wind it in as I had to stop the boat to get a better look at the problem. The drum had separated and by sticking my fingers through the gap I found shards of what I assumed were ball bearings from the seal. Damn it! I lay on my back and looked up at the problem. I could see where the two parts of the bearing had come apart. There were only three steel ball bearings left. The rest had disintegrated, presumably when I forced it earlier on.

I didn't have any spare parts for the furler and was at a loss to what I could do. It was an integral piece of equipment, enabling me to quickly and effortlessly pull the sail in and out, and it was virtually broken. I could still turn it to furl the sail up, but instead of pivoting on the bearing, the forestay would rub on the inside of the tubing which would eventually wear it away, rendering it unusable.

It was the early hours of the morning back in Melbourne so I waited until later that night, when it was about 8 a.m. at home, and gave Dad a call. It had been a long time since I'd spoken to him, so he must have been a bit shocked at what I had to report.

I explained the situation and asked him to contact the Australian distributors in Perth for a recommendation on what I could do. I said I'd call in four hours' time to see what he had come up with. Meanwhile it was dark and I was plodding along under mainsail alone. I had trouble staying awake as I waited to call Dad. When I did he had some grim news. There was nothing I could do to fix the problem with the equipment I had on board, which left me with two options: I could continue using the furler, risking the chance of other breakages, then pull into South Africa for a spare part and forego the unassisted component of my trip. Or take the entire unit off and alter the genoa so I could raise and lower it the conventional manual way.

I didn't like the idea of sailing with a faulty piece of equipment and South Africa was still six weeks away. I decided to get rid of the thing altogether and keep it simple. I had a new challenge—getting the bastard off!

The wind started to lighten as I went to bed thinking about my big job the following day. By the morning there was only a slight breeze with small waves lapping the side of *Lionheart*, perfect conditions to get the job done. The sun was out and I was feeling positive and looking forward to the challenge. I started by altering the sail. As I had to take the genoa off the furler track, it was a good opportunity to use a new sail, so I got the spare genoa out of storage. This was a job in itself, unpacking and packing again then getting the big bundle up into the cockpit. This spare sail had been made for the furler

Please, God, Stop this for Me: Azores to Cape of Good Hope

and didn't have hanks attached, which held the sail to the forestay and allowed it to slide up and down. I had about five hanks in my sail repair kit, so I had to take some off the spare storm sail and no. 3 jib. I had enough to cover the length of the sail.

I started work on the new genoa. I counted how many hanks I had and measured the length of the sail, then worked out how far apart each one should go and marked the spots with a texta. Then I used some sticky-back sail cloth to reinforce the areas and sewed a lip of webbing where the hanks would go through. This took me several hours. The next step was to cut the holes at the designated intervals, then put the hanks in and bend them shut again without snapping the brass gates off totally. I was relieved to have finished that job, but I knew it was the easier of my tasks for the day. On to the hard part. How the hell was I going to get that furler off?

I'd called the distributor in Perth earlier in the day to discuss the procedure, but I hadn't seen the guts of the thing yet and wasn't really sure what I'd find inside. The first step was to get the existing sail off the track leading up the forestay, which I did by unfurling the sail, letting it flap in the wind while I let the halyard go and pulled the sail down off the track. Half of it fell in the water but I was able to get it up on deck to begin the monstrous job of folding it up as tightly as possible, which was not easy in such a small space.

It wasn't perfect but, after 45 minutes of scrambling about, it was folded. I then organised my tools and started to unbolt the unit. I took the different-sized screws and bolts out and put them in a Tupperware container until I was able to pull both sides of the drum off and disassemble the section holding the bearing. The main mechanism was taken off, leaving the task of getting the tubing off the forestay. This proved the most arduous job of all, as I had to undo the forestay which held up the mast from the bow. I slackened off the backstays to take the pressure off, then slowly unscrewed the base of the forestay. With the small waves rocking *Lionheart*, the swaying of the mast was magnified, sending the aluminium tube along the forestay snaking like a fireman's hose out of control. I undid the last of the thread holding the forestay and bear-hugged the end as I used all my concentration to hold the jerking end. The last thing I needed was to have it flick into the air and knock me unconscious. I grabbed the allen key from my mouth as I reached up to where the first section of tubing was joined and undid the grub screw ever so slowly. Once I had the filthy screw and allen key in the safety of my mouth again, I started to slide the first 2-metre section off, grunting as I went. I was not sure if I was doing it the right way. It got halfway off the end of the forestay then it wouldn't move. Not one bit.

I yelled at it. 'Get off you stupid thing! Don't you dare stuff up on me now, look at the position I'm in!'

No matter what I said, it still wouldn't move. I was having a hard enough time holding on to the slippery, greased end of the tubing and wearing the skin off my hands, let alone having to get the section of tubing off. I rested my forehead on my arm which was going up and down with the jerking and cried in frustration. I simply didn't know what I was going to do. I had no answer this time. For every question Mum had asked, I always came back with something that sounded good, even if she didn't know what I was talking about. But this time I'd been caught out. I just didn't know what to do next!

I had a bit of a cry then, and started to hate the furler with a passion. My emotions were so strong I decided to set my own path. I analysed the situation and came up with what I hoped would be the solution, which required a piece of rope

Please, God, Stop this for Me: Azores to Cape of Good Hope

and a hacksaw. I was still gripping tightly to the forestay to prevent it flying off, and I needed to tie it down so I could get the tools. With one hand I slowly undid a piece of rope tied to the lifelines. The knot seemed to go on forever. Amid my frustration, the allen key slipped from my mouth and fell towards the water. I heard it clink as it hit the middle rail of the pulpit and luckily it bounced back onto the deck. It would have been a disaster to lose that, as it was the only allen key I had that would fit those grub screws.

Once I'd released the rope, I secured the flying forestay and tubing to the bow so I didn't have to hold it any more. I was free. I got up and stretched my back and muscles and steadied myself on the railing while the blood momentarily drained from my head. I grabbed the hacksaw and cut open the tube to see why it had jammed on the forestay. I soon discovered a series of plastic separators in the tube that had caused the problem. They were promptly removed with pliers. I undid the rope and, with one end of the tubing sinking through the water, continued removing the tubing. It was hard work and I didn't want to get over-confident, but as each piece of tubing came off and the next one lowered, the forestay became lighter and easier to handle.

Finally, with all of them off and all the stays tightened again, I hanked on the new sail and raised it for the first time.

WOOO—bloody—HOOOOOOOO! Yeah! The job had taken me all day.

Once moving again I sat back and marvelled at the shape of the new sail and watched the water pass by. I was exhausted but buzzing after slaying the dragon that seemed too big for me only a couple of hours before.

I'd been tested and I'd pulled through. The feeling was unbeatable. I got straight on the phone to Dad.

'Guess what? It's fixed!'

I reckon he was glad to hear that. I was bloody glad saying it.

It was another of my highs for the trip. I was happiest and most appreciative when it had been blowing a gale and was starting to die down. It may have been 30 knots but I appreciated the fact it could only get better as opposed to being 20 knots and getting worse. Plus, I felt I'd really achieved something by making it through the gale. This was how I felt after tackling the furler problem. There was nothing special or unusual (apart from the obvious) about the day, yet I knew I'd achieved something which made all the difference. I felt worthwhile and had everything in the world to be excited about. It was all in my head, I know, but on a trip like that, all I had was my mind working with what was going on around me.

The southeasterlies kicked in, which really wasn't the ideal wind, as southeast was the direction of South Africa, where I wanted to go, forcing me to sail as close to the wind as I could. I headed in a south– southwesterly direction, which wasn't really the ideal direction, as I was actually getting further away from where I wanted to be. A high-pressure system that sat in the South Atlantic, known as the South Atlantic High, was sending winds from the direction I wanted to go, pushing me further west, back the way I'd come, up the coast of South America. They were the trade winds I'd loved so much on my way up the Atlantic but going into them was uncomfortable compared to the previous weeks of easy, downwind sailing.

The boat was handling it fine but I spent a lot of time in the cabin because of the water that came off the bow when ploughing into the waves and spraying all over the cockpit.

Because of the increased amount of seawater landing on the boat and the constant sunshine, the amount of salt on deck

Please, God, Stop this for Me: Azores to Cape of Good Hope

also increased dramatically. It was everywhere, forcing me to rinse my hands with precious fresh water before opening up the computer or the video camera for fear of contaminating the electronics.

One night as *Lionheart* sat becalmed, I stumbled out of bed and clambered up the steps to stick my head outside to check for any shipping lights. As I scanned the horizon I was startled to see a black bird sitting on the top of the outside instrument panel. He wasn't taking much notice of me so I decided to say something to him.

'Oy, ay, you,' I said, but he only looked at me, turned and faced forward again. He was less than a metre away from me yet he wasn't scared by my voice. I reached out, expecting him to fly off. Not that I minded him being there, but I was curious as to why he wasn't scared. I reached out but he stayed where he was. I took one step further and touched his wing. With a slight flutter he turned and faced the other way then, with a hop and a flap of his wings, he jumped on the lifelines and settled his feathers again beside three of his mates who I hadn't noticed until then. The thought suddenly hit me. I was more than 500 miles from land. Could this possibly be a bird that hadn't had any contact with people, so hadn't found a reason to be scared of them yet? I could just imagine him having flown the oceans all his life. I was stoked.

My hair had started to annoy me since the wind picked up. It was disgusting, with clumps of thick salt-riddled hair stuck together. It would blow in my face, tickling my nose and blocking my view. Something had to be done. I could cut it off and look even more of a dero or I could tie it back. As all my caps had already blown overboard, I decided to make a head band to keep the growth under control. I tackled my new task with gusto and soon enough I was proudly wearing my new piece of clothing. It made all the difference, and I wore it every day until my return, only taking it off when I slept.

> *Saturday, June 26*
> Just passed the course I took on the way up over two months ago. Strange to think I've already been here! Wind is still on the nose and I'm getting quite close to South America. Roger says a bit further and it should free up a bit. Hope so.

> *Tuesday, June 29*
> The ETA for home should be the 8th (earliest) to the 28th (latest) of September.

A week later things started to change. I was approaching Trindade Island, off the coast of Brazil, in the same direction that the winds were making me go. I was 200 miles away from Trindade when I got the detailed chart out and had a close look at my course. The way the winds were blowing, I'd pass to the west of the island in about three days and I was happy about the distance I'd remain from it. Over the next two days the wind freed up a little and I could point more in the direction I wanted to go. This brought my estimated route considerably closer to the island, but still far enough away not to be concerned about crashing into it. When you have been in the middle of the ocean for so long, you become nervous about sailing within 100 miles of land.

The following day, on 7 July, I was listening to The Doors while fiddling with the switchboard that had been playing up, when I came up on deck and there, off the port bow, was a high volcanic island just sitting there as it had done, no doubt, for thousands of years. I was blown away. I had to

Please, God, Stop this for Me: Azores to Cape of Good Hope

reshuffle my mental filing cabinet as this was totally unexpected. I thought I was far enough away that I'd be well out of view of the island. I expected my next sighting of land would be some part of Australia. But no, this tiny 3-mile-long island, which I'd been trying to avoid for the last three days, had popped up into view and changed all that. Talk about a shock.

Subject: Message from Inmarsat-C mobile
Date: Fri, 9 Jul 1999 16:55:29 GMT
From: *Lionheart*
To: Katie Head

Dear Katie,
Well I was going to surprise you with a phone call cos the costs are down to $4\frac{1}{2}$ bucks a minute (instead of 15) but I couldn't find the envelope with your # on it. So could you tell me? It won't be a surprise now but then again you won't know when I'm gonna call so maybe it will, sort of. My ETA for home is between the 8th and the 28th of Sep, but at the rate I'm going now, it could be much longer. Hopefully I can catch up in the Indian Ocean.

TNT
Jesse

Subject: ring ring!
Date: Mon, 12 Jul 1999 13:45:29 GMT
From: Katie Head
To: *Lionheart*

Hey Jesse,
Well you are sounding happy at the moment! That's so cool about the cost of phone calls!

But try to call somehow when it is night time here otherwise I won't be home and not on weekends OK! That's if you still want to call!

Your ETA is smack bang through the holidays! Hopefully you get back for the snow!

My hair at the moment is in plaits all over my head! Very cool and very easy!

Well I got to go as it's late and I have to go back to school tomorrow! But I have Wednesday off! YAY!!

Speak to you soon.
Katie

I was using the Iridium phone more because of the drop in costs. I think I ended up speaking to Katie for half an hour! Email was still my main form of communication because it gave me time to write what I was thinking without wasting time and money trying to think of what to say on the phone. Plus, I couldn't afford to call all my friends.

I got the news that my website had received 400,000 hits since it had come on line a few weeks after I started. That's 17,650 separate visitors, which is amazing. Many of those were school kids, as the address was on the bottom of my *Herald Sun* column. They could email questions to me, which I received weekly, and made me laugh every time I got them. Here's a sample of some of the questions.

Please, God, Stop this for Me: Azores to Cape of Good Hope

Where are you heading?
Are you all rite? (sic)
Have you come close to any icebergs yet?
Will you have time to email us?
Have you been seasick?
What do you do in your spare time?
Have you seen any sea animals?
We want to know if you are having fish for lunch today? We will arrange home delivery of some chips to you if you are.
How many in your family?
What is the largest wave so far?
What is the best meal you have had so far? (Cheddar cheese on Vita Weats with an iced coffee milk.)
I can't wait for you to get back because my Dad has promised me the day off school to see you return. (Unfortunately I came back on a Sunday, so I presume he had to go to school the next day.)
How will the Easter Bunny find you?
Do you have a hammock or a bed?
What footy team do you go for?
Have you nearly been swept overboard yet?
What temperature will it be when you arrive at the equator?
Do you ever wish you could go back in time and stop yourself from going on this trip?
Have you seen any messages in a bottle?
Are you getting any money out of this?
How does it feel to be the only person in the world not to have seen *The Phantom Menace*?

I tried to answer all the questions I received, but as the PR company was getting up to 400 questions a week, it became hard. But every question was very welcome, and gave me reason to think about why I was doing the trip.

I continued steadily south continually monitoring my course on the chart as the wind allowed me to sail my chosen path, until I was only 20 degrees off heading straight for the Cape of Good Hope, South Africa. This was the fourth of the southern capes I had to round on my trip.

From this point on, the air took on a new kind of smell. It began to have the feel of the Southern Ocean. It was noticeably colder, bringing back memories of the first leg of my journey, which seemed so long ago. I started seeing albatrosses again, which I'd missed in the tropics and it was time to get out some warmer clothes and have my wet weather gear on standby again. I then realised I'd done something really stupid.

When heading up the South Atlantic all those months ago I dried out my jackets and overalls in the warm weather then packed them in garbage bags which were stored in the bow of *Lionheart* with the food and rubbish. But I couldn't find them anywhere. I re-searched the black rubbish bags up the front and checked all the containers.

What started as an incredulous suspicion grew to full-blown realisation. I'd thrown the bag containing the clothes to Mum at the Azores, thinking it was rubbish. Of course, Mum didn't check through my rubbish, and threw it straight into a dumpmaster. So much for $2000 worth of really good Musto Offshore gear! I was left with one set of Musto trousers and two sets of snow gear, given to me by a company from Western Australia. They were better than nothing, but certainly didn't have the protection of the Mustos. I was really mad because it had been such a waste, and it was all my fault. I still had a long way to

Please, God, Stop this for Me: Azores to Cape of Good Hope

go at 40°S in the middle of winter before I got home. Damn it!

I was between 30° and 35°S when the wind changed to come from the southwest. This was another milestone, for this was the westerly airflow that would take me all the way to my doorstep at the heads of Port Phillip Bay. I was glad to have the wind coming from behind once again.

> *Thursday, July 22*
> It's back! And Roger says this is it for sure, the start of the westerlies. A lot more albatrosses flying around, which gives me the feeling of returning to my long lost home, even though it has only been a few months.

I had rounded the corner and was in the home straight. I reckoned I could nearly see the finish line up ahead. There were no more bends to make or bits of land to dodge, except for South Africa, but from that point the course would hardly alter.

I continued on my way on a course taking me just south of South Africa. I still had two-thirds of the South Atlantic to cross because of the awkward course I was forced to take when heading south but that didn't matter, as I had the wind and currents with me. I was moving very well compared to the beating into the wind I'd been doing while heading south.

The aim was to pass South Africa at 40–41°S to leave at least 300 miles between land at my closest point. It was certainly not a case of skirting around this cape, like it was with Cape Horn, as the Cape of Good Hope sat at only 34°.

The Cape of Good Hope may be the best known cape around that area but it was not the furthest south. That honour belonged to Cape Agulhas, the southernmost tip of South Africa. Directly south of this was the Agulhas Plateau, a huge underwater shelf many miles offshore. The water here was quite shallow, at about 770 metres, compared to 5–6 kilometres depth in the open ocean, and could prove to be a major pain in the arse when the weather got bad. Down the east side of Africa flowed a current of up to 5 knots. When this hits a strong wind coming from the opposite direction in the shallow water of the Agulhas Plateau, all hell could break loose. It could produce high and steep waves, rather than the big, long waves of deep water. These steep waves can split a tanker in two when the bow and stern rest on the crest of separate waves, leaving the middle unsupported and causing it to break in half. However, a small vessel such as *Lionheart* can zip up and down the big waves like a cork.

In much the same way as Cape Horn, the African land mass also squeezed weather pressure systems into tight spaces, causing high winds to whip around the Cape of Good Hope.

I had another knockdown, the first since the Falkland Islands. It had been blowing around the 35-knot mark during the day but *Lionheart* had been handling the conditions well and the waves were starting to subside. It was probably 25 knots and the breaking waves were becoming less and less frequent. I was confident the conditions were abating so I went to bed content that my only job that night would be to get up to put more sail up as the wind died down.

A couple of hours later I found myself lying on the wall beside my bunk in the middle of a knockdown. *Lionheart* came back up, stalled for a couple of seconds, then picked up momentum again and went along as if nothing had happened.

I got up, annoyed that all my possessions were soaking wet and in places they shouldn't have been. It made me feel worse because the conditions weren't that bad and were dying down. It must have been a freak wave that caught me in the wrong place at the wrong time. I did a quick tidy up then went outside

Please, God, Stop this for Me: Azores to Cape of Good Hope

without getting into wet weather gear because of the generally tame conditions. I slid the slide back and pulled out the washboards then stepped up and outside into the cockpit. I had the first washboard in my hand and was in the process of putting it back when a second freak wave hit. It got me on a better angle than the first and ploughed over the stern, sending water flying down below into the cabin and soaking me.

I angrily put the washboards back in place and shut the slide while abusing the wave that had soaked me. I'd just got out of bed and was dripping wet with a stiff wind numbing my skin. I was not happy.

My first aim was to keep the boat moving with the waves in case of any more large ones that could knock me down again if I was caught side-on. I wasn't too concerned because the conditions were relatively sedate. I went back to the wind vane to set it on course again, only to find the wooden vane had snapped off, forcing me to get another one. Once I'd screwed it in place again, I set the tiller up and started tweaking the lines until it could correct itself without my help.

I tightened the tension on some lines that had been knocked by the first wave, then went back to bed in a soggy cabin. If it had been terribly rough then I could accept the knockdown, but because this wasn't the case, I found it hard not to be annoyed, simply because I wasn't expecting it to happen. Isn't it strange how our expectations change the way we look at things!

I had a Walkman radio with me which wasn't much use in the middle of nowhere as there were no radio signals to pick up, but I could play tapes on it. I had a few tapes that friends had given me, including two tapes of seafaring songs including the aptly titled 'Six Months in a Leaky Boat' by Split Enz, a tape of old English speaking which came with my literature books and a tape my friends Sarah and Carolyn had put together. I played this one over and over again while standing outside in the cold thinking about how home wasn't all that far away. I could really taste the closeness.

A few days after my last knockdown I was doing the rounds of all the shrouds and stays, tightening them up. The last time I'd done it was just after the Azores. When I got to the port diagonal shroud, from the mast spreader to the stern, I didn't like what I saw.

Where the fitting that connects to the deck was joined to the rigging wire was discoloured with orange rust. Most of the shrouds had rust on them, but this was different. Three of the outside strands of wire had snapped off completely.

It was obviously still holding, but when it came under maximum pressure I didn't have complete faith that it would hold. If one stay let go, that put more pressure on the others, which could follow suit. It was something I'd been fearing, and trying the whole trip to avoid by minimising stress every way I could. The only solution was to replace the shroud.

The job of replacing it didn't concern me. It was pretty straightforward. What did concern me was whether other stays would soon do the same because I only had a limited number of spare wires and fittings. I was comforted by one thought. Maybe the knockdown a few nights before had placed extra stress on the inner forestay carrying the no. 3 jib which was out the starboard side.

With the glimmer of hope that I could blame the knockdown and that it wasn't something I could expect to happen to all the shrouds, I got my equipment ready. The conditions had died down and it was a good day to get the job done.

First, I had to slacken the tension on the opposite side of the mast and undo the faulty shroud from the bottom, leaving it loosely attached. Then came the job of climbing the mast. I'd pulled all sails down and *Lionheart* was bobbing about, side-on to the waves. I started to climb the mast, clipping my safety harness to each of the steps as I got higher. By the time I reached the top, the sway of the boat was magnified so much that as the boat went from one side to the other, I was out over the water. I bear-hugged the mast until the rocking slowed to a manageable rate that I could work with. I pulled out the split pin and undid the second larger pin that held the shroud to the mast, then threw the wire overboard to ensure it wouldn't hit the deck and crack the fibreglass.

Back on deck, I measured the old shroud against my new wire, and cut it to length, struggling with bolt cutters for fifteen minutes to do the job. I screwed the norseman fittings to each end and connected my new shroud to the deck. The moment of truth had arrived—would it fit? Had I measured correctly? I was very worried, because I only had enough wire left to replace another shroud. I climbed the mast again and attached it at the top, then came down and tightened it at the bottom. It was a perfect fit. When it was done I was again relieved I'd overcome something that could have proven to be a major problem.

> *Sunday, July 25*
> I've taken up the act of inhaling the smoke from a smouldering match. Cos the sea air is so pure I don't get much variation in the odour department. It was when I started using matches for the first time to light the stove that I got a whiff of smoke which brought back memories of land and campfires and trees. I now indulge in this little reminder each time I light the stove. It brings back memories— a bit like looking at photographs.

While in the Southern Ocean, I'd use the stove not only for cooking but as a heater to keep the cabin warm. I'd taken plenty of metho, enabling me to light the stove nearly every morning. When I was outside during a squall getting a reef into the sail quickly I'd often get quite wet, so the stove was nearly always draped with sopping clothes. The leaks in the cabin worsened throughout the trip until I had three tea-towels on permanent drip duty, wiping drips before they landed on my computer or the back of my neck as I worked at the navigation table. It was a constant battle. I'd squeeze the soaked tea-towel onto the floor, which was home to a permanent puddle from my boots and clothing. When the wind was up or it was raining, most things down below were wet. It needed at least two days of sunshine or no waves crashing over the boat for the floor to dry out and the tea-towels to become stiff and dry. There was never a moment when something wasn't drying out!

I began to get some rumblings from home. As I had to take the long detour to the west of the high-pressure system sitting in the South Atlantic, there was a growing concern that I wasn't going to be home by 6 October, which meant I'd miss David Dicks' age record. I think some people thought I was deliberately sailing a safe route at a comfortable pace. It really got to me. The most important thing to me was that I got home without doing any damage to the boat. I could see the wear and tear on *Lionheart*, and if there were any more breakages then I'd possibly be forced to go to South Africa. I felt compelled to fire off an email to vent my feelings.

Friday, July 30
Everyone seems to think I'm about to give up or something. If I don't get back before the age

Please, God, Stop this for Me: Azores to Cape of Good Hope

record, then it won't really bother me as people might expect. For example, Kay Cottee carries on about how her trip was her life's dream. That's very well but I don't feel the same. Not that I don't appreciate the opportunities I've had. I consider myself very fortunate in every respect but every project in my life will be the most important thing at that particular time. I'm already dreaming about the next one. I've got almost everything I've wanted out of this. I'm not going to give up though, it's the last thought on my mind. I was a bit annoyed at the situation the other day with the rogue wave, but I have to wear it and do what has to be done. I'm trying to get home as fast as possible to see everyone, but I'm making a conscious decision not to push the boat just to get back 'in time'. I've made it this far by taking it easy. I'm big on principles and I feel if I compromise my principles now, then my success in future exploits or even the rest of the trip will be compromised. I value my conservative principles more than the record.

So I appreciate the support and I'd like to assure you that everything is under control. I should make it home before Oct. 6 but if I don't I'll blame the wind.

I think I made myself pretty clear. There was not much I could do but continue on my way, and get home when nature let me.

Subject: Message from Inmarsat-C mobile
Date: Wed, 4 Aug 1999 14:13:51 GMT

Lionheart

From: *Lionheart*
To: Trav Heenan

Hi ya Travo,
I've worked out that in 10 days I should be under South Africa and be on the final run home across the Indian. There is truly nothing new to say.
Been fine-tuning a song on the guitar, fine-tuning my pancake flipping skills and sleeping. A couple of electrics have stopped working and as today is quite calm, I'll get into the back of the switchboard and see if I can figure it out. ETA still the end of Sep. So get that beer ready!

TNT
JM

I was getting across the South Atlantic in true Southern Ocean style—quickly.

I was also battling a leak in the cabin where the running backstays had been added to the boat before I left. That in itself was no big deal, except the water dripped into the electrical switchboard. I first noticed the leak after Cape Horn when I needed to repair some corroded wires. I'd been trying to halt the leak with Sikaflex, but to no avail.

It had already claimed some victims. The first casualty was the wind instruments which told me wind speed and direction. I could live without those. Next, my tiller pilot wasn't getting any power. But when, in a few days time, I tried to put some music on and nothing happened it was definitely time to do something about it. I discovered more wire in the switchboard had corroded, shorting the appliances. I cleaned and rejoined

Please, God, Stop this for Me: Azores to Cape of Good Hope

the wires before treating them with anti-corrosion paste and hoped it would get me home.

> Subject: great!
> Date: Fri, 6 Aug 1999 17:12:07
> From: Trav Heenan
> To: *Lionheart*
>
> MM,
> GreatnewsYourSoClose!BoyzHavBinDownThePro m LAstWeek.
> SawASeal+ASnak.AwesumTime.WillHav2CumW en UGetBac
>
> C yaMate,
> TH

Because Trav had his email direct rather than going through Barbara, he had to foot the bill, so he came up with short cuts like this.

I was only about a week away from the Cape of Good Hope when Roger slowed me down, as a set of fronts around the area were causing trouble. They soon swung in front of me, so Roger told me to get to the Cape as quickly as possible, round it and get out.

It was kind of exciting heading towards an area I knew had many people living in a big city. With Cape Town only about a week's sailing away, it was comforting to know I could be on land in a week if I wanted to. On the other hand, it made me feel good that it was near, yet I had the power to sail right past with only the finish line to focus on.

Soon I was three days away from Cape Town, the closest I

came to it, before I changed course a little further south to miss the Agulhas Plateau and minimise the chance of encountering bad weather.

> Subject: Wx 246
> Date: Fri, 13 Aug 1999 03:46:33
> From: Roger Badham
> To: *Lionheart*
>
> It looks like you have a series of fronts heading towards you. Best to stay about 40. Further south looks like it's worse but you don't want to be further north either. Best to stay where you are or a bit further south.
>
> | Date | Pos | Wx |
> |------|-------|--------------------------|
> | 13/8 | 40/14 | WNW–WSW / 25–35kts |
> | 14/8 | 40/17 | SW–W / 15–25 kts |
> | 15/8 | 41/19 | WNW–SW / 25–35 (40?) kts |
> | 16/8 | 40/22 | SSW–WSW / 25 kts |
>
> Roger

I knew it could be a wet and wearing week as I passed under the horn of Africa so I got everything ready. I gathered food that was easy to prepare and stored it in the galley for ready use, set out my clothing and made sure the sheets to the storm sail were in place. The storm sail was already hanked on and I also attached the halyard to it and tied it down with a line ready to release when needed. Finally, I tied the genoa to the starboard life lines.

The first front came through the next day, an extension of

Please, God, Stop this for Me: Azores to Cape of Good Hope

the stiff wind I was enduring at the time. I grabbed the anchor rope, tied knots in it for resistance, and dragged it from the back of the boat to help keep *Lionheart* on course by making her a little more sluggish on broaching waves. There was little else to do, except spend many hours stuck down below on standby in my wet weather gear, while standing up, strapped in place by the galley life line and reading *Lord of the Rings*.

The wind died down as the first front passed during the night and it started to rain, a sign that the front was coming to an end. The rain calmed the breaking waves down until they were still large but more fun as they didn't break.

The next morning was relatively warm. I was still rugged up but the air and wind didn't have the same bite it usually had. It built up again during the day, bringing warm air from the land mass north of me, as I was directly under South Africa. This front was more of a concern as it not only brought the smell of land and a humid feel, but produced large thunder clouds that shot daggers of lightning into the water as they passed by. It was dark, so I stayed outside to keep an eye on the menacing clouds. When *Lionheart* was in the boat yard being prepared I put a grounding plate through the hull in case of a lightning strike. But even that was no guarantee I'd be safe if I was hit. The only way to find out would be to take a strike, which could smash through the fibreglass if that meant a shorter route to the water. I stood on deck watching the clouds and hoping like hell they wouldn't come near me. I felt like I was in a movie with that beautiful Enya song 'Storm in Africa' going around in my head.

I had the deck light on as I stood watching the storm. It shed some of its light out over the deck and onto the water passing by. Off to starboard I noticed something white sitting in the water. As the boat got closer I realised what it was.

Sitting there, with its wings tucked away, riding out the bad weather was a big albatross. It was uncanny. It just sat there as I passed by and disappeared into the night behind me. I wondered if he was as surprised as I was.

The weather died down a little during the night and the cold came back again. I was really looking forward to the end of the fronts. At least I was making good distance and getting out of this danger area, even though the conditions were pretty yuk.

By lunchtime of the 15th, the winds had picked up again but this time they were peaking at 40 knots and the waves were getting bigger and breaking more often with greater force. I was surfing down the waves and travelling too quickly to keep the boat under control. It was time to get the proper drogue out, rather than use the anchor line that was still out the back.

The wind was getting up to 45 knots and I was having to hand-steer with the drogue trailing behind. The cockpit was often swamped by white water that would tumble in from the crests of waves, drenching me.

It also scared the hell out of me. I started to get the feeling that something wasn't right. The wind was stronger than Roger had predicted and was increasing at an uncanny rate. This had happened only once before on the trip, just past New Zealand when I hit the force eight gale. I figured an unpredictable low had formed on top of me in only a few hours. Roger had no way of knowing it was going to happen. I could only hope the winds wouldn't get any stronger.

I was getting very cold in the cockpit but I couldn't go down below as the wind vane couldn't handle the steering required to keep *Lionheart* heading downwind with the waves. It was starting to get dark and I found myself in a conundrum. I was cold and it was getting dark and the way things were looking,

Please, God, Stop this for Me: Azores to Cape of Good Hope

I really didn't want to be outside in the dark for fear of waves washing me overboard or knocking me down. On the other hand, I needed to be outside to hand-steer and keep on a safe angle with the waves.

I was stuck. I knew what I wanted to do, which was go below and go to sleep to forget the terrible weather I was in. It was definitely the strongest wind I'd come across and the waves were getting BIG. It was hard to estimate how big they were but when the biggest of them came along, I knew a small yacht shouldn't be out there with them. I'd say they were 10 metres high, from trough to crest, but it was hard to say. They could easily have been more or less, but it didn't really matter as I was starting to get worried!

Was this going to keep increasing and turn into a hurricane? Not knowing what was ahead was the scary part. At the time *Lionheart* was handling them quite well. I was getting wet and thrown about and there was a real danger of being thrown overboard, but apart from those 'maybes', I was doing well. Sort of.

I was still trying to decide what to do when the answer was given to me. *Lionheart* started picking up speed and I sensed something different about how the tiller responded. I looked behind and saw that the line attached to the drogue was limp. The drogue line had snapped, which meant I had no brake. I faced a terrible battle to steer *Lionheart* with the waves, and I'd surely get hypothermia outside all night, so I decided to drop sail and sit out the storm in the cabin by laying a hull.

As soon as I stopped the boat by turning side-on to the waves and tying the tiller off, the noise dropped and the boat didn't rock nearly as much. It was quite a comfortable position to be in except that the wind blowing on the port side pushed the boat on quite an angle to starboard. I tidied up outside then checked for any breaking waves before opening the hatch and going down below.

It wasn't all that nice in the cabin as it was darker and I was dripping wet and shivering. I lit the stove to warm up and got a tea-towel, which was still wet, and wiped the excess water off my face and hair. After I'd warmed up a little and hung my jacket up to drip, I got out of my overalls and boots, leaving me in thermal underwear and wet socks. I hung the socks up over the stove and turned the flame out. I had to do this because often it would blow out and leave the metho dripping until it overflowed onto the floor. I wiped my wet feet on the outside of the lee sheet and stepped into the bunk and under the sleeping bag. It was moist from the salt and the drips above the port holes but after a while my body heat dried it out. I flicked the light off above my head then lay still listening to the boat and the waves as I tried to get to sleep and prayed that it would die down.

Some time during the night I was woken by something I'd been expecting. It was a knockdown—a proper knockdown. It was dark and there were things flying about the cabin, with water that seemed to reach every corner of the boat. As *Lionheart* came back up with the sounds of water spraying from the roof area, I flicked on the light to find I was on top of my sleeping bag with two Tupperware containers underneath me, a couple of books and my guitar resting where my head used to be. Everything was wet! My soaking sleeping bag reflected the light as it glistened with water. Everything was so wet that I thought a porthole must have been smashed by the force of the wave. I stepped over the guitar and flicked the electric bilge pump on, as well as the second light above the navigation table.

'Oh man, I HATE this,' I said out loud as I steadied myself and started throwing the loose objects floating around the floor up into the V berth to clear the area. I was still dumbfounded

Please, God, Stop this for Me: Azores to Cape of Good Hope

as to where the water had come from. There were no broken portholes and everything else seemed in place.

I was taking in the mess when the second wave hit. It was as powerful as the first and I watched in astonishment as I saw it happen for the very first time with the lights on. Thankfully I was holding onto the vertical grab rail leading to the roof from the navigation table, and quickly secured my other hand on the handle leading out into the cockpit. The boat went over as I hung there. I held myself by putting my feet on the side of the galley cupboard while in front of me the navigation table opened up, spilling the charts, binoculars and other bits and pieces. *Lionheart* went well past 90 degrees—I estimate it would have been as much as 120 degrees. I felt the whole boat vibrating under the strains of the wave and I soon discovered where the water was getting in. Above my head came a stream of water like that from a faulty shower rose. It was icy cold and proceeded to drench me as the pressure of it sprayed out everywhere. Aha! It was the tiny gap between the hatch slide that was letting the pressurised water from outside get in, not a porthole.

I waited and waited for what seemed like an eternity but was in fact only a matter of seconds until the lead in *Lionheart*'s keel brought me back up again.

Two knockdowns in five minutes. Was this the start of a sequence? The wind was howling outside and a glance at the wind speed instrument showed 50 knots, gusting to 55. It had got worse. Who knew when it was going to stop? I could very well expect more knockdowns so there was no way I was going outside with those freak waves coming in like that. The boat could get swamped as I was getting out or, worse, I could lose one of the clumsy washboards overboard when putting them back in place. That open space would be disastrous considering the amount of water that came in through a tiny gap in the slide.

I expected another knockdown soon. I didn't know whether the mast could handle the power of those knockdowns that had pushed it well into the water. I also thought of the shrouds I'd repaired only a few weeks before. Had I been careful enough during those calm days in the tropics when the flapping sail would send vibrations throughout the rig? Had I been pushing it too hard in the name of speed? I certainly hoped not but there was nothing I could do—except pray for the conditions to die down, which was precisely what I did.

My bedding was wet inside and out, including my mattress and pillow. I too was wet and shivering, so I stripped down to nothing and winced as I got under the cold, wet sleeping bag and flicked out the last light above my head. I rolled myself up into a ball with my head under the cover and breathed against my skin to warm up. I was all on my own in a situation that I'd no control over—I started praying.

It became a chant as I repeated it over and over to show my sincerity.

Boom! Another one!

I didn't bother looking, I just heard the water from the slide as it landed in the puddle that had formed on the floor.

'Please God, stop this for me. Please, just make the weather get better, that's all I want. I don't want to go home. I mean, it'd be great to see everyone but I want to get there myself. Just make the weather better so I can do that, get there myself because when it's good out here, it's beautiful and I love it.'

I was woken another two times that night by knockdowns but they weren't as severe as the initial three, with more time between them. My suspicions had been correct. A small low had developed near me and caused the wind to get stronger than predicted. These lows form quickly and they also move quickly, so by the next morning there was a considerable difference.

Please, God, Stop this for Me: Azores to Cape of Good Hope

It was light and sunny when I got up, confident that I could get the boat moving again. There was still a 25-knot wind blowing, but it seemed like an afternoon breeze compared to the previous night. My first look through the clear washboards revealed that the starboard solar panel, the side that was getting knocked over, had disappeared. I wondered what else was missing so I checked through the skylight and was glad to see the mast still standing.

My first priority was to get the cabin relatively comfortable and that meant drying everything. The stove wouldn't start. I changed the metho but it still didn't work. It had been thrown off its hinges and had landed upside down, spilling metho, the previous night. It took me half the day to find the blockage then drill it out and fill the hole before I could begin the huge task of drying things out.

A few hours later, once the cabin was nearing how it looked before the storm, I went outside and was shocked by what I saw. The genoa was hanging overboard to starboard with the forward stanchion snapped off and the lifelines drooping down. The pulpit had also taken a beating. The stainless steel frame had been bent right over to port until the starboard side frame was only an inch from the forestay. The genoa strapped to the lifelines must have caught the waves, bending the stanchion until it snapped and bent the pulpit. The starboard spinnaker pole had also disappeared while the pole on the port side sat there with nothing holding it in place.

First I got the genoa back on board and tied it to the port lifelines, careful not to go overboard on the starboard side as there were no lifelines. I then started the job of getting the snapped stanchion off the lifelines to cut it in the vice down below, then bend it round again and fit it back a little shorter than what it was before.

[243]

There was nothing I could do to bend the pulpit back into shape so I just left it and reattached the lifelines and re-tightened them. They were a bit wonky but would still do the job. The solar panel that had gone was one that didn't work any more anyway and the wind generator gave me enough power so I wasn't fazed at the loss.

There was one more front forecast to come through in a day's time then it looked clear for a while.

I'd made it through a force ten storm with a bit of damage but nevertheless I was through and racing to get out of the area before any others came about. I was so relieved to have calmer weather. But the challenges were far from over.

CHAPTER 9

The Final Run: Cape of Good Hope to Australia

No power.

I stood on the deck scratching my head. It was three days after the storm and I'd finally got things in order, but then my power cut out for the first time on the trip. This was not a simple case of some corroded wires preventing me from getting a fix of Ben Harper. Not one single electrical thing would turn on. A check of the meter revealed a problem—down to only 7 volts. It was meant to read a healthy 12 volts, as it had done for eight months. I had no lights, no radar, no email and no radio.

The most frustrating thing was that the wind generator was spinning like mad in the bluster of the Indian Ocean, but wouldn't charge a thing. For some reason the power drained from the batteries, even without anything turned on. The only power on-board was in the satellite phone. Its battery was almost empty, to the extent it showed no bars on the battery meter. I figured I may have a few minutes of talk time, if that.

I slowly worked through all the symptoms during that day before giving Dad a very quick call at night. I drew a deep breath as the phone rang. When I heard him on the end of the line, I shot out, 'Hi Dad, it's me. I've got a problem with the batteries and this phone battery is nearly dead. It's down to 7 volts and the wind generator won't charge. It spins but no amps are going into the batteries. There is a bit of corrosion along the line of the genny but there's a lot of corrosion all over the place. I'll call you in two hours, can you ask Richard what my best option would be—change to the spare wind generator, check all the wiring or something else? Also, please tell Mum that I'm fine but I can't send email. Speak to you later. Thanks.'

I hit the off button and took another deep breath. Dad had contributed the words 'Hello' and 'Bye' to the conversation.

I felt a bit sorry for Dad. I didn't speak to him that often but when I did I usually had something to say that he probably didn't want to hear. I couldn't begin to imagine what he was thinking after the abrupt call. It was the middle of the night when I made the call, as I was eight hours behind Melbourne. I'd timed the call to catch Dad at the start of the day, so he could make a few calls for me. He was able to get onto Richard, the electrician who had worked on *Lionheart* and fitted the switchboard before I left. I suspected the corrosion on the line near the generator was the cause of the problem, but I was not about to start repair work until I heard what Richard had to say.

There was nothing I could do except keep sailing until Dad called back. At least I didn't need any power for that! It was a fair night so I made good progress.

Dad's two-hour deadline soon passed, so I put in the call. He was ready to go, with notes of what he had to say in front of him before the phone cut. Richard believed there may have been a short in the generator line, draining the power and, at the same time, not allowing the generator to charge. It could be the corroded area I'd already identified, or it could be something

The Final Run: Cape of Good Hope to Australia

blown inside the actual unit. Then again, it could be a problem anywhere. I thanked Dad, then went to sleep that night planning how I'd tackle the problem the next day.

As soon as my eyes opened I saw a bright, blue sky through the portholes. I leapt out of the bunk and into the day's challenges. It was amazing how I could fire up when I had something to do. These projects came along so rarely. My first job was to put the brake on the wind generator to stop it from sending a charge down the wires as I handled them. I grabbed my tools and took them to the lazarette, the hatch on the back seat of the cockpit, and unpacked all the spare ropes, buckets and rags.

The corrosion was around a join in the line that led under the deck. Some malleable tape had been wound around the join which theoretically should have sealed it from the salt. Something must have gone wrong. I unwound the tape and pulled the two ends apart. There was black and green muck everywhere. I spent ten minutes cleaning the strands of wire with a copper brush then wiped them with soldering acid to clean them properly. With this done, I wound the ends around each other and began the fiddly job of soldering them together. Once I was finished, I wrapped new tape around the exposed area and sealed it. A hop, step and jump down below and I hit the buttons on both the battery meter and wind generator brake. A wave of relief washed over me as I watched the amps increase on the little digital read-out as the generator built up speed.

I called straightaway on what was left of the phone battery and told Dad that it was only a short causing the problem, and not something more complicated. It was such a relief, not only for me and him but also for Mum, who didn't know what was going on. I waited until lunchtime when the batteries had enough volts to run the DC adaptor for the phone and gave her a call. She was so glad to hear from me. It was one of the worst moments of the trip for her, as the unknown loomed large in her mind, even if it was only for one day. I think of how David Dicks was out of contact for ten days at one stage. Imagine how that would be for a mother.

I'd just had two of the biggest frights of the trip in less than a week: a full-blown storm with five knockdowns and the threat of no power. I felt as though I'd had it all by now. I was still only past the bottom of South Africa, but really looking forward to getting home. I'd always looked forward to my return, of course, but beforehand I was experiencing new things all the time. I was still enjoying it out here but I felt I'd come through the final test and I wanted to get home without any more weather challenges or equipment failure. Luckily, I had only the Indian Ocean between me and the people I wanted to see at home.

> *Wednesday, August 25*
> The finish seems so close yet it seems to be dragging on forever. I've actually numbered the amount of days left to go and cross one off every day.

> *Wednesday, August 25*
> Well, something just happened how I never expected it to. I was outside changing the genoa when I came back down into the cabin and found an email waiting. It was Mum congratulating and telling me at the same time that I was now eighteen years old—I would have forgotten about it!

It was actually only fifteen minutes past midnight at home when Mum sent the email, so it was still the day before my birthday where I was. It was a bit confusing.

The Final Run: Cape of Good Hope to Australia

I'd originally planned to be home by my birthday, but that wasn't to be the case. It made me more determined to pick up speed knowing that I could have been home if I'd been travelling at the speed that David Dicks had been doing.

As for the specialness of the day, there was none. It felt like any other day. I was reminded by the guys on the television show *The Panel* that I could vote and do other things, but out at sea it meant nothing. It was another example of something I'd realised—you can be in paradise with everything you ever wanted but it's not the same unless you have good company. I spoke to my friends over the phone who had congregated at my place for some chips and drinks but that was more of a media set-up than anything else, though it was great speaking to them and Mum. I got another email the following day from Mum. She'd just heard on the news that a tornado had hit South Africa causing chaos and doing millions of dollars of damage. Who knows what could have happened if I'd been in that area at the time?

> *Sunday August 29*
> Fifteen knots of wind, moving along at 4 knots with the sun glistening off the water. What could be better, you ask? Well, how about if I was going in the right direction! For the last four days I've had easterly winds making progress very slow as I can only point NNE or SSE because the wind is coming from where I want to go. The result is that my velocity towards Melbourne has only been 80 miles—that's a measly 20 miles a day.
> To add insult to injury, I finished my last bag of food earlier this week and have been living off damper instead of the leftover dregs like peanuts

and muesli. A few quick calculations tell me that about two weeks before I arrive home, all the flour will have run out forcing me to eat the stuff I don't like. Even more reason to get home quick.

It was between South Africa and Madagascar when my good progress stopped. I'd been moving along quite nicely, covering more than 100 miles a day, then I hit five days of light head winds from the east. All I could do was head a bit further north after having drifted down to the south since South Africa. There was nothing to do but turn my attention to other matters.

I was getting into a big damper-making phase of the trip. Instead of mixing the flour with sugar and milk and making pancakes, I just made a dough with water, flattened it out and cooked it thickly on the frying pan. I did this nearly all day every day during those five days. It was kind of frustrating not moving much but not as bad as it had been in the past. I had the memories of a force ten storm freshly embedded in my mind. This was a minor irritation compared to that.

Thursday, September 2
It was about two days ago, just after a squall had come through, that I was looking out over the water lost in thought. I'm sure this isn't any new philosophical theory or anything, but I was thinking about the similarities between events in life and a rain squall. When it's approaching it looks pretty mean, dark and gloomy. Then, all of a sudden, it hits with a torrential downpour sending you off course and out of control. You

The Final Run: Cape of Good Hope to Australia

can't see anything around except mist and rain, but it doesn't last forever. Eventually, the first rays of sunlight seep through the tail end of the cloud. The light refracting from the tiny droplets of water suspended in mid-air put on a show that is truly spectacular. At this point you are the closest you'll ever be to heaven on earth.

I saw one of the most spectacular shows of nature that day. Out to starboard was the most brilliantly coloured rainbow that I had ever seen. The dark clouds in the background made each and every colour stand out as if it were alive with electricity and it was so close that I could see the end fading into the water only about 30 metres away. I concluded that the heavy pot of gold must have sunk, but I didn't mind because I felt invincible to the passing squall, or any other nasty weather that this world could throw at me. The thing is, had I not encountered the squall, then I would have missed that feeling of jubilation. It made me think that bad times are just preparing the way for better things to come.

Sunday, September 5
Still not moving very well—maybe I shouldn't have caught and cooked that albatross! (Only kidding) My position is 38°41′S, 48°14′E. It's been another week of terrible progress! In the last sixteen days I've only averaged 50 miles towards Melbourne per day. I'm supposed to be in the strong westerly air flow but I've had varying, easterlies, and squally weather making constant movement in the right direction very hard.

The wind had picked up, which enabled me to turn my attention to my goal—the right-hand edge of the chart. When I got to that point I could then turn it over and marvel at the beauty of the other side, for on this side, up in the top right-hand corner, was a little yellow shape that cut across the corner of the page. This bit of land was Western Australia. I was not far from home. It's funny, but to think of sailing from Perth to Melbourne, a distance of about 1300 miles, seems like a long trip now that I'm home, yet there I was, looking at WA on the map and considering myself nearly there. But there were a few creases and folds in the map I had to sail over before I got there, including passing very close to a strange little island, Amsterdam Island, which sat on the Mid-Indian Ridge, a bit over halfway between South Africa and Australia. When I passed that I knew I was really on the home straight.

> Sunday, September 12
> In a few days I expect to be 3600 miles away (roughly) with an ETA of 36 days.

The good weather continued. Of course there were ups and downs, times when it was wet and cold outside, but that mattered little, for *Lionheart* was heading home. This was the longest stretch of continuous movement for the entire trip.

I'd hit the fabled Roaring Forties. This was the area around latitude 40°S, where the wind roared around the bottom of the globe, pushing a sailboat along at good speeds. The old clipper cargo ships could do the England to Australia passage in fewer than ten weeks by using these winds.

Even with the movement, which I loved, life was pretty uncomfortable as I crossed the Indian Ocean. Ever since water had poured into the cabin during the storm off South Africa,

The Final Run: Cape of Good Hope to Australia

it had not dried out properly. In fact, it hadn't been dry for months, not since the tropics. When I used the stove to dry the tea-towels or clothes, the steam would stay trapped in the cabin. It would fog up the portholes and collect on the roof, then run down the hull and into the bilge. Or it would form into water drops and wet everything again. The process just went round and round. But the worst thing was the mould. It started to grow on everything—my sleeping bag, clothes and cushions. The cabin ceiling became covered in it, and it grew wildly in the dark corners.

The temperature was also dropping. It was very much winter in that part of the world. The passing fronts came from the south, sending down icy-cold hail showers. The air temperature hurt during the fronts. I had no way of knowing how cold it was, as I'd broken my temperature gauge during the knockdowns near the Cape of Good Hope. All I knew was that I could only afford to be outside for five minutes with bare hands before I had to go back down and stick them over the stove.

I may have been getting closer to home, but I still had to be careful, as I was thousands of miles from land. That was reinforced one sunny day after I made a sail change. I jumped back down below to escape the bitter cold, putting the washboards back in place behind me. I then slid the hatch back in place. Something caught in the washboards, making them sit up slightly higher. The hatch jammed with the top of the washboard. I could see what happened, so tried to slide the hatch open again. It wouldn't budge. I tried again. Nothing. I tried everything I could think of but nothing would work—I was trapped inside. I'd often wondered about getting locked outside. I tortured myself with the thought of freezing to death or dying of starvation, but I never imagined it could happen the other way! Then I remembered the bow hatch, which I never used. I got my harness on and scrambled over bits of equipment and bags of food as I made my way forward. I opened the hatch and pulled myself up and out during a gap in the waves which were often spraying over the boat. I was careful not to let the hatch right down so the latch could catch, just in case I couldn't open the slide from outside (I'm sounding like a poet).

I made my way around to the cockpit and stood before the jammed slide hatch. With more room and a powerful leg, the heel of my boot connected with the edge of the slide and it slid forward with a crack. I went back down below and cleared the washboards so it wouldn't happen again, then went and closed the forward hatch.

On one of the few calm days of the crossing, I heard some strange squawking coming from up on deck. I immediately thought the wind generator had claimed a casualty, so braced myself for the worst as I stuck my head out. But the noise was coming from the foredeck. I spun my head around and there was a big albatross stuck in the lifelines up the front. He was in a mad panic as he awkwardly tried to free himself, tripping over onto his beak while his large feet made a racket on the deck. He must have tried to land on the boat or lifelines and misjudged his approach. I was not sure what to do. I expect he'd have been a bit bad-tempered and I didn't fancy having an albatross taking a swipe at me. However, before I had to do anything, he broke free and stood on the foredeck. But not for long. With one leap he jumped overboard in a half-hearted effort to fly off, landed in the water and paddled away like a scared cat running from a car.

Saturday, September 18
Had some strong winds recently and haven't made the best progress. Another 33 days to go.

The Final Run: Cape of Good Hope to Australia

The middle of the Indian Ocean is a hell of a windy place. It might not get the strongest bursts of wind but on average it beats Cape Horn for its constant winds. The stretch around Amsterdam Island was where Roger told me a lot of the Around the World racers set their speed records because of the fronts surging one after the other. But for a boat like *Lionheart*, which I'd been desperately trying to nurse home since the start of the trip, those fronts would not mean record speeds. They actually held me up, as I could not afford to race with the wind. I was forced to drop sail and hove to more than at any other stage of the trip. The fronts lasted for three weeks, longer than Roger originally predicted. I comforted myself with the thought that this would be the last of the dangerous conditions for the trip. I figured that when I got across this section things would get back to normal and the chance I'd get home without any more knockdowns would increase dramatically.

Amsterdam Island got closer as the strong winds pushed me along very well, until I could flip the chart. And there was Australia up in the corner. I was nearly home.

> *Friday, September 24*
> Just got in contact with Perth Radio for the first time ever. I tried about a week ago but with no response so I tried again, hoping the distance I'd travelled since then would help. It did. They sounded pretty happy to hear from me as well and I got the frequencies for Radio Australia off them, so I'm just about to program that in and listen to some more Aussie accents.

It was another milestone that represented how close I was to home. I'd been used to the BBC, Spanish from the Atlantic radio traffic and the heavy accent with Raphael in the Canary Islands, so an Australian accent was like a long-lost language to me. I programmed several frequencies into the radio and one of them worked immediately.

I had to change between the three depending on what time of the day it was but this wasn't a problem. I had a new toy to play with and though I usually hated watching cricket at home, I sat and listened to it as if it was the most beautiful thing I'd ever heard because it represented life on land—the life that I'd be soon re-entering.

Mum had been in frequent contact with Pat Dicks throughout my trip, and on a recent chat she learnt that David was leaving on a delivery yacht from Fremantle, heading for Sydney. She told me the name of the yacht and its call sign, so I asked Perth Radio to let me know when it started to take position reports so I could tee up a sked to have a few words with David. Things were starting to get busy and I was enjoying it!

Tuesday, October 5
Just heard a noise up at the wind generator. It was raining but I opened the hatch slide and there was a small bird that looked like a pigeon sitting on the seat. There was blood splattered over the seat cos he had just flown into the genny. I was going to get out and help him but he saw me first and flew off into the water. I saw him try to fly off but he got a metre or two into the air and came back down into the water. No doubt a shark will get to him, the poor bugger.

I had expected this would happen. I was such a novelty to passing birds they all wanted to come have a look at me, and

The Final Run: Cape of Good Hope to Australia

were fascinated by the humming of the generator. And with the constant spinning blades, the fact it had taken nearly ten months for a bird to fly into the generator was pretty amazing. Luckily, for the collision caused considerable damage, not only to the bird but to the blades of the generator. I flicked on the electric brake to halt the terrible clunking noise it made.

One of the blades had snapped in half while another had its tip lopped off. I decided to replace all the three blades as each needed to be the same length for the thing to work. I stopped the boat, as I always did when making repairs, then stood looking at the propeller, working out how I could safely position myself to unscrew the unit. I found I could sit on top of the solar panel with my legs around the pole holding the generator and make my repairs. As always when you need to climb a height into an awkward position, you always forget something. I was up and down the frame up to ten times, trying to find the right-sized tools. The freezing cold conditions affected my work as I clumsily fiddled with the screws to put the unit together. I was glad to get back into the relative warmth of the cabin. I was happier when I flicked the brake off and the generator purred into action.

> *Wednesday, October 6*
> I'm really into the mood of returning. Got Radio Australia on and catching up on all the news. ETA 18 days to go. Current position is 40°29′S, 106°18′E. I'll continue along the 40°S line, which will take me under Albany which will be 300 miles away at the closest point.
>
> Looks like I'm arriving just in time for the good weather as well—forecast looks generally better over the next week and I keep asking

Lionheart

myself If I've had the last blow or are there still more in store.

I'd been at sea for exactly ten months, which forced me to take stock. When I left I expected to be away for nine months, so I gave myself a month's grace and packed food for ten months. Basic mathematics told me I had another three weeks until I got home. Commonsense told me I had a food shortage problem on my hands.

Throughout the trip I'd been going into the next day's food bag and stealing the lollies and good things to eat. The stuff I didn't like I packed in rubbish bags and stored it up the front. It was now time to go back through the bags and select the not-so-bad-any-more items to live off.

On my first pick of the slimy plastic bags, I selected all the mug noodle packets which I'd earlier gone off. I threw them up to the galley where I wiped them clean and dry with my trusty tea-towels. They were then stored in a new garbage bag in the galley. These only lasted a week, so I was forced to go back again and select the next best things. The selection process underwent another two rounds before I got home.

In the last few days I was eating freeze-dried mushroom pilaf which, if left overnight, resembled the consistency of canned dog food (but not as appetising), stale cereal without milk (I'd run out), peanuts and Musashi energy cakes.

Actually, I started to love Musashi cakes. They smelt better than they tasted but this wasn't a problem. They were the only type of wheat substance I had, and they soon became a favourite.

I was making good progress towards Western Australia, and started looking at the next chart covering the Great Australian Bight to Port Phillip Bay. I was starting to get some pressure for

The Final Run: Cape of Good Hope to Australia

an arrival date from home, particularly from Barbara, who had to get things ready and organise the media. The Yacht Club also needed to know so they could get things ready as well.

It had been decided as far back as May, about the midway mark, that I'd arrive on a Sunday as that would give people a greater chance to get down to Sandringham to greet me. I thought it would be best to come in on the weekend, although I didn't like playing up to it. But I had a responsibility to my sponsors, particularly Mistral and Sandringham Yacht Club, to come in on a day that suited them. I was also conscious that a lot of school children were following the trip through the pages of the *Herald Sun*, so this gave them a chance to come down as well.

But, having said that, I was still uncomfortable with what was planned. I did not want a big fuss made. As far as I was concerned, a fuss was just what was being planned. My progress indicated I had two weekends to choose from—Sunday 24 or 31 October. To make one could mean a mad dash and a heart-stopping finale. To make the other may mean wallowing around Bass Strait to stage an entry that may have looked too contrived. I wanted to be back on the earlier date, but I was starting to have strong doubts.

Sixth October was the date I needed to return by to break the outright age record. It passed with little fanfare as I had known for weeks I wouldn't be back by then.

> *Friday, October 8, 2.16 p.m.*
> I don't think I'll make it back by the 24th. I've got another sixteen theoretical days to go but I really need to be at the Heads the day before to get everything ready and so forth. Possibly the Sunday after.

For the entire voyage I hadn't come close to being washed overboard. Despite it being one of the fears of everyone involved in the trip, it had not presented itself, even when I was caught outside in my first bad weather off New Zealand. That was until this point. I was aware that I couldn't become complacent just because I was nearing the end. I'd made it as far as I had because I'd been very strict and careful in the way I conducted myself and how I sailed *Lionheart*. But what happened had nothing to do with complacency. It was just bad timing, which could have happened at any stage of the voyage.

I was standing on deck, halfway up the boat, with the wind hovering around 25 knots. My harness was clipped on at the shrouds as I re-attached the spinnaker pole on the port side which continued to come undone. The swell was average but it was just bad timing that when a gust of about 40 knots came, *Lionheart* was on the face of an extra large swell. She took off down the wave and turned side-on and began to tip on her side, plunging the port lifelines halfway under the breaking white water at precisely the point I stood. I was plunged into the swirling water, totally at the mercy of the sea. I couldn't fight it, so I just waited and wondered to what degree the boat would go over. It paused in a leaning position while I hung on grimly as the currents washed against me and the freezing water shot up my sleeves. It seemed like ages, but was only a few seconds until *Lionheart* came back up. I was straddling the lifelines, half on the boat and half off as *Lionheart* shook herself like a dog that had just taken a dip before continuing as if nothing had happened.

I knew I'd had a close call. If I was to fall off the boat, even with the harness on, there was no-one to help me back on again. But I wasn't thinking about that. The incident happened

The Final Run: Cape of Good Hope to Australia

so quickly and everything went back under control so soon that my only thought was to blame the stupid spinnaker pole whose fault it all was!

> *Saturday, October 9, 3.35 p.m.*
> I'm nearly under Perth. Current position is
> 39°41'S, 111°47'E.

David Dicks left with his delivery crew just before I passed under the first longitude of Australian soil. They were on a much larger ketch that was capable of faster speeds than *Lionheart*. Perth Radio set up a sked, so for several nights I was able to swap a few stories and comparisons with him, despite the patchy radio reception. We chatted about technical stuff and experiences we'd both shared. It was good to talk to someone who had the same experience as me, who understood what I was thinking, and knew of all the little things that made up a trip like mine.

They were a little way ahead of me and had copped the worst of a low-pressure system that hit Perth, giving them 50-knot winds and wetting their boat down below. I was thankfully unaffected. If I was at the start of the trip I'd probably have been a little disappointed to have missed out on the storm. But I had nothing to prove to myself any longer.

> *Sunday, October 10, 3.49 p.m.*
> I just got out the last chart of the trip which shows Melbourne on the far right. I'm really starting to feel that I'm on home territory once again even if I'm not yet in Australian waters.

Lionheart

I'd completed all my requirements for a global circumnavigation, passing under the fifth and final cape, Cape Leeuwin, off the southwestern tip of Western Australia.

Monday, October 11
The decision has been made for me to come in on the 30th. I'm not going to make the 24th.

Wednesday, October 13
Don't know where Australian waters start. I'm now about 250 miles south of Albany at position 38°43'S, 116°49'E. It seems like it is taking forever to get home.

Thursday, October 14
The wind has been doing strange things. Totally different from what Roger says the models show it to be. He has warned me that anything could happen cos of the strange weather systems about at the moment.

The strange weather in the Bight brought about a fair few days of fog. During one of those days, with rain squalls constantly passing by, I just happened to stick my head outside for some fresh air. Lucky I did, because I noticed between the swell the hull of a large ship heading north, and on a possible collision course with me. I ran downstairs and called on the VHF radio to see if they'd spotted me. I wasted a couple of minutes trying to raise them, but with no response. I quickly went up into the cockpit to see the ship was getting closer. I gybed the triple-reefed mainsail across and changed direction to starboard, eventually getting out of the ship's course. I sat and watched

The Final Run: Cape of Good Hope to Australia

the steel monster pass close by. There was no-one on deck, and they presumably hadn't seen me. It was the first sign of human presence I'd seen in ages, and soon they disappeared, having disrupted my day.

The excitement was really growing, and the quiet lifestyle I'd been leading was left behind forever. I felt as if the trip was already over. I knew I could sail around the world and I was living in harmony with my surrounds. I didn't feel like I had to cross the finish line to prove anything to myself—I already knew it.

Sunday, October 17
Moving quite well. Expect to have four days up my sleeve waiting outside the Heads if I continue at this pace. I'm only 900 miles away from Melbourne and I still can't make radio contact with Sandringham Yacht Club. Every time I try I get some guy as clear as a bell saying in an Indian accent, 'So you want to come into Bombay?'

Monday, October 18, 11.58 a.m.
I've been cleaning the mould off the roof and the slime from the deck to make *Lionheart* at least half presentable for when I get in.

Tuesday, October 19
I've got 800 miles to go from today—that means theoretically I'll be there on the 25th but I've got some head winds coming up which could slow me down. Over the last ten days I've averaged 75 miles a day but that includes two days of no wind. We are still working out what time I'll be

coming through on the 31st. It could either be 1
a.m. or about 7 a.m.—Mum is finding out which
way daylight savings go to see if there is enough
time left to get to SYC if I come in during daylight
(7 a.m., preferable).

I was making my way across the Great Australian Bight, yet I felt as though I was out on the Bay, because home felt so close. I really needed to concentrate, as there was so much going on around me about my arrival, and the growing media attention, that it could have been quite easy to take my mind off trying to get *Lionheart* and myself home without assistance.

Wednesday, October 20
Yesterday I had the fishing line out and hooked
an albatross but he got off himself before I pulled
the line in.

Friday, October 22, 10.52 a.m.
Wind is being a pain. I'm at 39°10'S, 134°20'E.

The conditions were much better than what I had across the mid-Indian, but it was still very much wintry conditions and not the best weather one could hope for.

The excitement grew as I got busier and busier each day, taking media calls and keeping an eye on the forecasts to work out when I'd arrive. I was enjoying each day tremendously. It was like the day before your birthday—whatever you do is enjoyable because you know that tomorrow is your birthday which you can always fall back on to make you feel good. It was like this for me for a whole week. I wasn't in any rush to get home because I was just enjoying the thought of it. That may sound

The Final Run: Cape of Good Hope to Australia

strange, because I desperately wanted to see my family and friends, but the anticipation of getting home was such an incredible feeling. It was similar to when I passed South Africa and knew there was a huge population not far away from me, but I kept going. Or even that feeling of anticipation when you're about to kiss a girl: the holding back, the power and discipline. I enjoyed having the choice and control and not rushing in.

I was one week from home. I was still about 200 miles off the coast of South Australia, but in my mind, I was home. Pretty much as soon as I knew the forecast for the week the danger had pretty much gone, as there was little chance of rough weather. It's strange, but the finish pretty much fizzled out. There was no great realisation that the trip was suddenly over. I cannot put my finger on a point and say that was when I realised I was going to make it home. The homecoming gradually built up, through increased contact with home, and more frequent media interviews. There were a number of points which signalled the end. The first was as far as the Cape of Good Hope, when I moved to a new chart, which had the west coast of Australia on it. Then, later on, I came in radio contact with Perth. But I suppose, if there was one defining moment, it occurred below Adelaide, when the *Herald Sun* sent a plane to take a photograph of me. The last plane I'd seen was the lights of a commercial flight in the North Atlantic Ocean. I'd been excited by a rock in the water (Cape Horn) and now a plane flying in the sky. Was this a sign I'd been away for too long?!

There had been a lot of build-up and preparation for the picture. The *Herald Sun* had been trying to organise pictures of me at various points throughout the trip, without success. The closest we got was at the Falkland Islands when I outran the RAF planes. The *Herald Sun* had been with me the whole way, and wanted to be the first to get the picture. (I have since learnt that to be the first is pretty much everything in the media.) After many calls over the weekend, a plane was dispatched from Melbourne on the Monday before my arrival. I was a bit sceptical of their chance of finding me let alone getting a good shot, as I had a 20-knot wind, which made for a pretty ugly sea.

But, at about 1 p.m., I heard a noise I hadn't heard for a very long time—an engine. Suddenly, with a whoosh, the twin-engine plane swept over me. It did about nine sweeps, the last coming so close I could see the faces of those on board quite clearly. They were the first faces I'd seen since the pretend pirates in the Atlantic. It felt great to see them. I then knew that if I did get into trouble, I could be reached.

I didn't realise at the time what that photograph would do. It was splashed across the front page of the paper, triggering an amazing reaction from other media. It seemed as though the starter's gun had been fired. Within days I was visited by all manner of planes and helicopters. The helicopters would hover right down low, sending *Lionheart* shooting off from their down draught.

> *Tuesday, October 26*
> A lot of crap over when I should come in. I'm coming in on Sunday morning but if the forecast is bad I'll come in the night before. I've decided that I need a codename so I can talk to Dad on the radio with some privacy. My codename will be *Imajica*, the name of Dave's boat, rather than Lionheart.

The Final Run: Cape of Good Hope to Australia

As I passed the Victorian coastline I was able to pick up FM radio for the first time. It was strange to hear the new music at the top of the charts and the advertisements that hadn't changed since the year before. It was stranger to hear them mention my name on radio and that I was off Portland and for people to keep an eye out. I couldn't see land yet, so I very much doubt they could have seen me!

What was stranger than hearing my own name on the radio was hearing my own dorky voice. I didn't realise I sounded like such a moron. I was listening to my new favourite—Radio Australia—and I caught a replay of an interview I'd done a day or two before. I'd like to blame it on the satellite phone but I'm afraid after so long without much talking, I sounded a bit strange!

The air traffic got busier as I got closer to Melbourne, but I was happy to see them, as it reassured me I was closer to home. I forget how many choppers came out in total but I remember that one of them told me they had a surprise and I heard the question, 'Have you got a song ready for us, Jess?'

It took me a few seconds to recognise the voice over the static VHF but then I realised it was Dad. Another television station, not to be outdone, brought Mum out to see me, hovering above in a helicopter while I talked to her. It was uncomfortable having a conversation with Mum that I knew would be broadcast on television. There was the growing number of radio and press interviews, and I needed to finish my columns for the *Herald Sun*. I was starting to feel like an animal in the zoo. At one stage I was being advised to avoid one television station because another wanted to do something with me. It all sounded a bit silly and I was a bit annoyed. Plus, how was I going to get away from a helicopter? The wind died down in the last few days, so there were not many big waves to hide behind.

On top of that, the planning for my arrival was well underway, even if I was convinced it was a lot of unnecessary fuss. I couldn't believe the trouble they were going to. For instance, Hayley White, who had sung at the AFL Grand Final, was going to sing the national anthem as I arrived. Then, after I was ashore and up in the clubhouse, the plan was for me to appear on the balcony and open a bottle of champagne. That was probably my biggest concern. It just didn't suit me to be spraying the crowd with champagne. That was something Grand Prix drivers do. I didn't see myself as that type of person. I just felt so stupid over all the fuss being made. I registered my protest with Barbara and Mum, but deep down I knew I had no choice. I'd just have to do it. I really wanted to turn up and see my family and mates, do the press conference, then go home and get on with my next project.

The final week was certainly not turning out as I had imagined. I thought I'd be sailing at a constant speed. I'd see the coast, cross the line, see the people at the yacht club and go home in the one day. A lot more pure, if you know what I mean. That sequence stretched across a few days. I actually slowed down to delay my arrival until Sunday, but then the wind died down in Bass Strait.

Over that last week, a debate began over when I should enter the Bay. There were only a few windows of opportunity for me to pass through the Heads. As they were so narrow—only a few miles apart—the force of the water as it surges in and out of the Bay on the tide can be horrendous. The fears were greater than when I left, as this was the point I'd finish the journey and the record attempt. Plus, I wouldn't have a motor to call on if I got into trouble.

John Hill had contacted Gordon Reid, an expert sailor with many years of experience guiding boats in and out of the Bay,

The Final Run: Cape of Good Hope to Australia

and I spoke to him a few times about what I should do. I wanted to come in on the ebb tide on the Sunday morning, and sail straight up to Sandringham. The worry, however, was with a crowd of people expected to greet me, including the new Premier of Victoria, I couldn't risk leaving my entry until the last minute.

Plus, as daylight saving began at 2 a.m., there was a sense that it would also add to the confusion. There was also some unsettled weather approaching, just to cloud the issue. John Hill urged me to come in Saturday morning at about 6.08 a.m., and spend the night inside the bay at Queenscliff. It made a lot of sense, but I had this niggly doubt in my mind that although the record attempt finished at the Heads, I'd be somehow cheating the people who came to watch. Plus, the media had been onto me for a week. Surely they would find me sitting in the Bay on Saturday afternoon?

I continued to enjoy the small things those last few days, like washing my hair. I was basically out of fresh water except for 30 litres I'd collected from the deck while in the doldrums. I grabbed one whole 20-litre jerry can and warmed up half of it over the stove then plunged my head into the bucket. No joke—I used about half a bottle of shampoo in one wash, which didn't take long to turn the water a muddy colour. Then a rinse and a comb with a fork and I was feeling like new, if not looking like a Lassie dog.

> *Thursday, October 28*
> The wind has died and I'm off the coast of King Island. Can't see it yet.

The water colour had changed from a deep, deep blue to a greenish colour as I got onto the shelf under Bass Strait. It was not long after that I noticed something more significant on the horizon—land. I'd thought about this moment for eleven months and it had finally happened. As *Lionheart* rolled up sideways on a wave, a thin strip of land was visible for a split second. I rushed outside and waited until I was on top of another wave and the horizon was clear of any other peaking waves and there, I saw it again . . . and again. It stretched a fair way across the horizon, and soon I could see a lighthouse at the northern end of the land. King Island!

I felt like I was on my back doorstep. I'd sailed those waters to King Island with Dad on *Bohemian* more than a year ago.

It was the Friday night when Roger told me that a front was possibly coming through. He said it wasn't looking too bad but the weather reports over the radio had it at up to 40 knots for the Saturday night then clearing slowly over Sunday. I continued to get a bit of pressure from home as everyone thought it was up to them to advise me when to come through. I knew that if it was too rough at the Heads when it was time to come through, then I'd just have to wait. There was nothing I could do. I could come through earlier on Saturday night when the winds weren't so bad but it would be dark then and probably more chance for something to go wrong. Roger maintained his usual laidback manner, saying that he couldn't see it being as bad as everyone else was forecasting. I decided to stick with my gut feeling and come in at first light on Sunday. I'd made a big call. But I continued to monitor the weather, and was ready to come in earlier if I had to.

There was little wind on Saturday as I sat in the cockpit 30 miles directly south of the Heads, inching closer. Another plane came out and took some photos of me playing guitar as I sat in the cockpit. Andrew also came out in a plane to film the boat from the air to add to the footage I was doing myself. I let

The Final Run: Cape of Good Hope to Australia

the fishing line out for something to do and before it was all the way out I had caught something. It was a fair-sized barracuda, which I threw back overboard.

On the horizon I could see a fleet of spinnakers coming through the Heads. They got closer and closer until a couple of them passed quite close. Some of them were actually from Sandringham Yacht Club, so they said hi and quickly shouted about the preparations going on at the Yacht Club. I was home for sure—the quiet days of the trip were over.

The wind died again in the afternoon as I was visited by the final chopper for the day. It was a quick interview because on the horizon loomed a mean-looking cloud that was coming in from the northwest. I knew the front was going to bring primarily westerly winds so I tried to get as far west as I could to save sailing directly into a headwind.

It was the scariest cloud I'd seen on the whole voyage. It was very low and moving fast in one great big long line and followed by rain and fog. I pulled all the sails down and waited on the flat sea for it to hit. It didn't take long to get to me but only reached 20 knots, pouring rain for only ten to fifteen minutes. Once again Roger had been spot on. I waited for the rain and wind to ease a little then got out and raised some sail and started heading for the Heads—this was truly my final run!

On Saturday night, in my mind, I was home. I was sitting less than twenty miles from the Port Phillip Heads, and I could see the lights of the towns dotting the Bellarine Peninsula. I could smell land for the first time in nearly eleven months. Strange as it sounds, I could smell grass, the bitumen of the roads and street lamps. Don't ask me what a street lamp smells like, as I can't describe it, but that was the picture in my head as I smelt that strange smell. I loved it!

I was out in Bass Strait, yet everything seemed so close.

Almost claustrophobic in fact. I was in a busy shipping lane, and tankers seemed to be skimming past only metres away, although they were quite some distance from me.

I have never felt that the world was such a small place as I did that night. I'd already discovered how small the world was, and not just by sailing around it. When you can receive a note from an eight-year-old girl in a small country school followed by a letter from Bill Clinton, President of the United States, and another from my idol, Ben Harper, you realise there really are no boundaries on this earth.

I had a bloody terrible night. The volume of shipping traffic required me to keep alert. I tacked back and forth, which was hard work, and I was getting more tired as the night dragged on. I was also trying to stay in a relatively small area to remain lined up for my run to the Bay.

Above all, I was excited. God, was I excited! But I was also apprehensive about what lay ahead. When I headed out of the heads on 7 December, there was a big unknown ahead of me. I had that exact same feeling as I prepared to re-enter the Bay. What would the next few weeks, months or even years bring?

I was also concerned about facing my family. I was unsure how I'd react, which made me uncomfortable. What would I say to them? How would I act? Would I hate being home? Would I be as normal as I thought I was? Would my friends treat me differently?

I had changed. But was that because of the time I'd spent away and my experiences, or the fact that I was eleven months older? I definitely had a lot of time to think about things, and I think I understood a lot more about issues. Things like how the world works, the dynamics of business deals, the workings of the media. And, possibly, I got close to understanding a lot about myself.

The Final Run: Cape of Good Hope to Australia

I sailed backwards and forwards, biding time until it started getting light. I'm not sure what time it was—it had to be a bit after 4 a.m.—when I finally grabbed the tiller and did what I'd wanted to do for a long time.

I headed home.

Chapter 10

Beyond the Waves

> So when storm-clouds come sailing across your
>   blue ocean
> Hold fast to your dreaming for all that you're
>   worth
> For as long as there's dreamers, there will always
>   be sailors
> Bringing back their bright treasures from the
>   corners of earth
>
> But to every sailor comes time to drop anchor
> Haul in the sails, and make the lines fast
> You deep water dreamer, your journey is over
> You're safe in the harbour at last You're safe in the
>   harbour at last.
>
> — 'Safe in the Harbour', Eric Bogle

The night sky turned a dark blue which slowly spread from the east until it began to lighten as early morning was born. I'd positioned myself well and the wind was a nice 10 to 12 knots, pushing me from behind and slightly to the side. The best time to come through the heads was about 6.45 a.m., so I was making good time. I knew Dad, Barbara, John and others would be heading out, all on different boats and all excited— but not as excited as I was. The time just flew as I stood at the stern of *Lionheart* for the last time. It was my favourite position to stand, protected by the solar panels with a good view of the boat as she sailed along. I had the VHF radio in hand to call the lighthouse keeper when I got close to entering the Bay.

The colours of the morning were amazing. Maybe it was just the rare beauty of land or something undefinable. The hills behind Sorrento to my right were a misty dark colour, with the sea a rich dark blue as the sun began to light up the clouds. It was magic. The headwaters appeared to be on their best behaviour. No mountains of waters or swirling whirlpools. Just a gentle welcoming swell.

I could gradually make out the lighthouse and the gap forming between the two points of land. I then saw the first of the boats. I didn't know how I should feel. This was the moment I'd been looking forward to for so long, yet I didn't know how to react. I was certainly excited because I'd soon be face to face with humans. Maybe the scariest part was how people would react to me. I'd never liked people looking at me, yet I was the reason those boats were coming out to the Heads.

So much happened from that point on. Things quickly became a blur and rolled into one. I called the Lonsdale lighthouse keeper as I neared the heads.

'Lonsdale lighthouse, this is *Lionheart*, can you see me yet? Over.' I asked.

'Roger that, *Lionheart*, yes I can,' he said, quickly adding, 'Can you get out of the way as there's some shipping coming through.'

I was taken aback. It made me feel small, like I was out of place.

I think the Coast Guard and Barbara's boat got out to me first, closely followed by other yachts and power boats, until more than a dozen crowded around. A couple of helicopters had also joined us by that stage.

We travelled along together until the lighthouse called with those magic words. 'All right, you've just crossed the line.' He told me it was 5.28 a.m., which didn't sound quite right. A few minutes later he called back with a correction—it was 6.28 a.m. when I crossed. It seemed I was not the only one a little confused by the change to daylight saving time!

I then saw Dad, and could see the relief on his face as he sailed beside me on *Bohemian* and we spoke over the VHF. I also saw his fiancée Suzie for the first time—there were new things everywhere.

There was no feeling of euphoria that I'd finished the trip, just a sense of relief. The nurturing of *Lionheart* and my approach of not pushing the boat had paid off. It had been worth the effort to pull down the sails as they jarred the rig in the doldrums. I was glad I'd stopped the boat during the gales and taken an extra day to get home rather than not at all.

I was relieved that I was no longer bound by the rules of the trip. From the moment I passed through the Heads, 328 days before, I had this thing hanging over me. It had been a strange situation out at sea—I had no rules about when to go to bed, when to get up, to do homework or any of the other thousands of rules that govern our lives. I was in control. Yet, I had this all-governing rule hanging over me that I had to do the trip non-stop and unassisted. There were times when I'd have loved to pull into Brazil, the Azores or Cape Town, but I simply couldn't. I was so free on one hand yet, on the other, I wasn't at all.

Beyond the Waves

Now I was free! I could do what I wanted—jump overboard, take someone on board. I could also finally receive things, which I gleefully did when I was handed a freshly cooked hamburger (which I'd requested), two pizzas, several cans of Coke and a packet of Tim Tam biscuits.

The realisation that I'd just sailed around the world never entered my mind.

I'd always been confident I would complete the trip, especially after the previous weeks of sightings, so crossing the line had no special meaning in a way. If I was constantly worried, or my equipment was failing, it may have been a different story, but to achieve what I wanted to do in the future, I knew the record was very important. As it turned out, I was three weeks older than David Dicks. However, I was able to complete the trip unassisted.

I dropped the sails to be towed the 40 miles towards the Sandringham Yacht Club. Everything for the day had been organised and I had to arrive there at a certain time. It was pretty much out of my control. I wanted to sail if I could, but at the time I just went with the flow.

In hindsight I could have sailed that leg easily, as I dropped the line about 10 miles from the Club three hours ahead of schedule. I was in a real bind. I couldn't just lob into the Club, as much as I wanted to. I ended up sailing in circles, surrounded by dozens of boats, and the ever-present helicopters. I'd just sailed around the world, but I was forced to sit 2 miles offshore before I was allowed home. I started to twig at the level of attention I was receiving. All around me were boats, battling what were not the most pleasant sailing conditions.

Sitting in the Bay while boats milled around me and people yelled greetings was a bizarre experience. I could see something along the shore line and cliffs surrounding the Club.

Only when I got closer, after being given the call to come in, did I realise that something was thousands of people who had come to see me.

I couldn't believe my eyes. I'd expected a big crowd, as David Dicks had come home to a large welcome, but I didn't expect the sea of people before me. Where the hell did they all come from? They reckon there were 25,000 people present. I had expected a big crowd at the Club, but what I didn't expect were the thousands clamouring for vantage spots on the cliffs as I sailed up the Bay. They were everywhere!

Sure, it was a bit unusual for a seventeen-year-old to take a couple of years off school, raise some money and sail around the world. But it had been done before, and someone will do it again. I had no understanding of the number of people who had followed my trip avidly, particularly my weekly diary in the newspaper.

Having been back for a while, I now think I understand why so many turned out that day. There was a storyline over eleven months, with ups and downs, and trivial and life-pondering moments. And it was real. For many people, it made a difference from the usual diet of television. For many, that Sunday afternoon was the climax of the story they'd been following for so long, and they could play a part in the ending.

I finally got the call to come in a bit before 1.30 p.m. Dad sailed the last mile alongside me. I quickly dropped the genoa and pull it on board as a Yacht Club dinghy drew alongside to manoeuvre me around the breakwater. There were people everywhere you could imagine, trying to get a look. The crowd started to stir until it turned into a cheer and applause. I didn't know what to do or how to handle it so I just focused on steering the boat and looking as if I was concentrating. I hadn't seen anyone for five months, and suddenly thousands were cheering me.

Beyond the Waves

As I got closer the crowd got louder and louder. I could then see the small jetty that stuck out on its own like a baby's finger. Standing on the jetty were some familiar faces—friends and family—and a tall dark-haired man named Steve Bracks who I later found out was the state's new Premier.

I couldn't wipe the smile from my face. I've never smiled so much in my life. Even if I wanted to I couldn't stop. At that point I believe I was experiencing genuine happiness like I'd never felt before. I was also suddenly filled with pride at what I'd done.

Then I saw Mum for the first time since the Azores. She'd chosen not to come down to the Heads, as she felt she would have been frustrated at seeing me but not being able to hug me. She was standing on the edge of the jetty. I could also see Oma, Gran, Beau and Andrew. Dad and Suzie were still behind me on *Bohemian*. I'd been homesick, angry, scared and frustrated, but when I saw my family, I felt something that I just can't describe.

The boat slowly glided into position next to the dock and I took time to make sure I was secure. Then something strange happened. I just sat there, as though not knowing what to do.

Someone said to me, 'Come on, you can get off now,' which jolted me out of my reverie.

I didn't know what I was doing or thinking. I've looked at the news footage since, and it was obvious that I was sitting there unsure of what to do next. I stood on the side of *Lionheart* and slowly lowered myself onto the jetty. I had no time to check the stability of my legs for Mum immediately grabbed me and jumped up and down. Her emotion washed over me.

After greetings all round, I took my first step. I felt like a clown, as though I'd had too much to drink. The mind knew what it wanted, but my feet were unable to carry out the instructions. It took a few seconds, but I began the walk to the Yacht Club. I turned to look at *Lionheart*. I wanted to see what she looked like, to see her beautiful lines from an angle I hadn't been able to see her from for such a long time.

I was soon overcome by the crushing crowd as I began the 150-metre walk to the Yacht Club. It was incredible. There were kids and dogs, and people pushing others as they tried to touch me—like the fervour you see in religious festivals in countries like India. It was made even more dramatic by the fact I was being given a wedgie by two security guards who had hold of me by the back of my pants.

There was definitely a lot more media at the Yacht Club than I expected, probably 25 reporters, plus associated technical people and cameramen. I believe there was even crew from the BBC. The questions were harmless enough, even if one of the reporters tried to quiz me on my views on the upcoming republic vote. I knew enough to stay clear of that one and tried to stick to my adventure. I think I gave him a sufficiently confusing answer that it never made it into print. Others asked if I'd found God, or if I took Playboy magazine. I answered 'yes' to one of them.

It was over in about fifteen minutes, then came the moment I'd complained about weeks before—greeting the crowd and opening the bottle of champagne. To tell you the truth, it wasn't that bad. I was going with the flow of what had been organised. Talking to the crowd made me realise that I'd been given a special opportunity and that I should use it wisely. I'd never had an audience like that before and probably never will again. There were old men, young girls, mothers, wild sons, every type of person imaginable. I tried to hammer home my theme that young people can do anything and should be encouraged to follow their dreams. I'd have liked to have been there as a spectator, to take it in properly, appreciate it more, and see it from a different angle.

Beyond the Waves

The official schedule for the day said my next engagement was a medical examination. It sounded so serious, but it consisted mainly of Dr Broomhall asking if I was all right, and my replying 'yes'.

The next step was something I'd been looking forward to for a long time: a shower and a change of clothes. It went all too quickly. Putting the clothes on gave me an indication of whether I'd lost any weight. Hardly any, as it turned out. All my clothes fitted perfectly. Actually, they were Beau's clothes, as he had trendier clothes than me.

Then came time to see my mates. They were the same bunch I left, and luckily, they treated me the same. They appeared to have taken advantage of a few free drinks at the bar, so they were carrying on a bit. After most of the crowd had left, about eight of us went down to the end of one of the jetties and took a leak off the end. Some habits were proving hard to break.

We left Sandringham at about 5 p.m. I travelled the hour back up to Sassafras in Mum's car with Mum, Andrew, Beau and one of his mates. We chatted about pretty mundane things. Nothing about the trip, just family life. When we arrived home, the media frenzy began.

There was a television crew waiting to film me entering the house. I was also filmed in the shower, as though it was the first since my return, but our hot water service wasn't strong enough to provide the steam for that 'realistic' shot. I stood there for 30 minutes while kettles were boiled to create the desired effect. It was a bit tiresome.

The media finally left late that night, which allowed me to do what I'd yearned to do for months. About 10 p.m. I went around to Darren's place, where I celebrated with all the boys (and a few girls) with a few beers and a sing-a-long until the early hours of the morning.

It was great to be back! I rolled home at 2 a.m., where I found Mum still up, so we chatted for an hour about the trip and what had happened at home in the last eleven months. I got to bed at 3 a.m., after about 40 hours on the go. My head hit the soft dry pillow in a bed that didn't rock.

And so ended the biggest day of my life.

~

The next day, the media frenzy began in earnest. I was up at 6 a.m. for a live interview, then spent the day in a farcical effort to protect one television show's exclusive rights to me. It got to the stage where I was told to cover my face while doing a radio interview so another camera crew couldn't film me. I was pretty annoyed at being put in such a situation. I knew all the media attention was warranted, that I was newsworthy for that moment, and it was valuable for me to build my profile for what I had planned, but I wish it could have been a bit more settled in those first few days.

But there were some fun times. I arrived a few days before the Melbourne Cup, so my family was invited along to watch the races.

I also received a civic reception at the Melbourne Town Hall, which was a great honour, and gave me my first chance to catch up with all those who had supported my trip.

The next few weeks were pretty hectic, with a lot of media interviews, particularly for some of the magazines, which had been arranged before I had come home.

~

Not long after my return I received a call from the Sandringham Yacht Club to say Prince Philip was visiting Australia in March and wanted to meet me! The visit had to remain a big secret until a few weeks before the visit. I kept up my end of the bargain, with only family knowing what was going on. Someone else obviously didn't feel bound to keep things secret—the news I was to meet Prince Philip made it onto the rumour segment of one of the highest-rating breakfast radio programs in Melbourne. There was great excitement at the club over the visit, given that Prince Philip was actually Commodore-in-Chief of Sandringham Yacht Club and had visited the club before.

I arrived on the day, and went down to where *Lionheart* sat, at the exact spot I'd arrived five months before, but not before I spoke to the media for a few minutes.

'Are you looking forward to meeting the prince?'

'Well, yeah, but I didn't lose any sleep over it last night.'

'Why, is there someone else you'd rather meet?'

'Well, of course, I'd be much more entertained by a super-model [giggle from the reporters] but yes, I am looking forward to it.'

Three white cars entered the yard and out jumped the Prince. A quick shake of the hands with a few of the heads of the club, and it was time to meet him. I committed my first faux pas for the day with my hand in my pocket as I greeted the Prince. I honestly didn't realise, as I was just standing comfortably. I found him to be a perfect gentleman and he was genuinely interested in looking at *Lionheart* and asking about my trip.

After a few minutes, he wandered off and the media pack descended on me again to ask how it all went. I told them what I felt. That it was nice to meet the Prince, and I was glad I had as it gave me a better understanding of how life must be for the royals, but it was not the highlight of my life. After all, the Prince was just a normal person, like me, in the same way that

I now understand that Bill Clinton and Ben Harper are only human.

Someone asked me if it was worth sailing around the world to meet the Prince. Of course it wasn't. What a silly question. I also added that I thought he'd gone to so many of those things that he'd probably not be too interested in everything he saw.

The two daily newspapers really got stuck into me the next day. The *Herald Sun* carried the headline 'Jesse a fish out of water'. The story told of my mismatched attire of blue blazer and black pants (I got dressed in the dark and didn't realise) and lack of excitement over meeting the Prince.

The *Age* told a similar tale, reporting that I was 'nonchalant': '"It was nothing special." He said the former navy officer knew his yachts but had declined an offer to board the *Lionheart*. Prince Philip was nice, he said, but it probably wasn't worth getting into a suit to meet him.' (The comment probably owes more to an indifference to fine tailoring than any personal response to Prince Philip.)

Mum was pretty upset by what was in the papers, and I believe the Yacht Club got a few calls over what was seen as my disrespect. It was my first encounter with negative media. I certainly said those things, but it was not what I meant. I thought the Prince was a great bloke, and I was very honoured to meet him. But the meeting was not the highlight of my life, and not the reason I sailed around the world.

The next day, amid all the worry over what people would think, the phone rang at home. It was the Duke's personal secretary, Mike Parker. He called to say that the Prince had a wonderful day and was pleased to meet me. He also said the Prince was aware of the controversy over the stories, and wanted to let me know that he knew what it was like to be misquoted and misrepresented by the media.

'Welcome to the club,' was his message.

But it made me think about how the media reacts when you get a profile. I was opening a solar boat race in Canberra some months after I returned, which involved my being winched down from a Navy Sea Hawk helicopter onto a solar boat. However, the down draught from the helicopter began to push the boat away from me. The chopper, with me dangling from the winch, had to chase the boat until two army guys onboard grabbed my legs and pulled me down. As far as I was concerned, it all went to plan. I thought the boat was meant to be moving and it was fun, if not a bit uncomfortable in the rescue sling.

The next day's front page of the main Canberra newspaper told how I nearly died while doing a stunt out of a chopper. It was disappointing to see them try to sell a few extra papers by making up a story rather than reporting on the positives of the boat race and the school kids who'd put days into getting their boats ready.

~

A few months after I returned I found myself busy doing talks all over the place. But I still had a debt of more than $80,000 hanging over my head, and I'd given up my pamphlet job ages ago. After a few months with no money coming in and the debts continuing to grow, I got in contact with Phil Gregory, whom I'd met and chatted with via email during the trip. I asked if he'd like to organise talks for me so I could earn an income.

We found that we had similar personal values and goals in life, and I related very well with him. He is basically my manager, although I hate to use that word, and our relationship isn't really like that. He'd just given up his position as an advertising agency director and hadn't done management work before, so we were both beginners. We now work together to tell others they can turn their dreams into reality.

What I value most is being able to use my public profile to share the lessons I've learnt from the trip. I don't particularly want to be remembered as a sailor. The trip was something I wanted to do and now I've done it. But I've got heaps of other goals, including telling people how powerful I believe the human spirit is, and inspiring people to have faith in themselves.

One of the ways I'm doing that is through the Reach Youth organisation. Reach was formed six years ago by AFL footballer Jim Stynes and film director Paul Currie to help young people from all walks of life who may be facing the challenges of being a teenager. These guys had no formal psychological qualifications, but by sharing themselves with compassion and speaking on a level that is real, not superficial or authoritative, Reach has achieved amazing results. I've been in rooms where the toughest kids you can imagine have been brought to tears, and consequently changed their view of life.

And Reach doesn't just assist those who go through its courses. I've found that its message has helped me since I came back. For instance, it encouraged self-expression to the extent that I've been able to write this book. I'd never revealed my feelings of rounding Cape Horn, my homesickness when I left or how scared I was in rough weather, until I sat down and put it on paper. This book is my way of telling the true story of my journey, to close that chapter on my life so I can get on with the next challenge.

It's my chance to tell people what it was like out there, to get into words all those things I find too hard to describe when I'm interrogated by curious people.

Beyond the Waves

Whenever I meet someone, inevitably the question comes up, 'So what was it really like?'

Or 'Weren't you lonely?'

Or 'I've done some travelling myself—where did you stop off along the way?'

I get asked such questions so often, I find myself reciting a well-worn response that seems to lack enthusiasm. It's just that I can't get enthused about answering a question when the person expects a quick response that sums everything up. I simply can't convey the beautiful sights and experiences in a few words. It's taken four months to put these thoughts into a book and, in many ways, writing this has been therapeutic.

One statement I hear over and over again that I totally disagree with is: 'I could never do that.' As if I'm some kind of freak capable of something beyond the realm of the ordinary human being. I'm just a normal person! I spend time with my mates and do all the normal, naughty things eighteen-year-olds do. My friends treat me the same as before I left on the trip, which I really cherish, because with them I can truly be myself.

Having met a few celebrities since I returned, including Prince Philip, I dislike the perception that people with public profiles are gods. What I did was an amazing achievement and I am very proud of it. But it stops there because I'm referring to the achievement, not me!

What pushed me to give it a go in the first place and then succeed was the human spirit that lies in all of us—the spirit of adventure. To activate this, however, we have to push the boundaries and have faith that we as humans will be able to overcome any hurdles in the way.

I have a theory that our lives are governed by some mysterious and unexplained laws, which once we have put in the effort, will lead us along the right path. That is precisely what

I'm proud of—heading out into the unknown in the first place and allowing those laws to take care of the rest. The hardest part is the thought of hitting the tarmac and, until you do, you're living through the toughest phase of it.

These laws are very much alive. Small miracles pop up everywhere. I was fortunate in a lot of things—that I got a great sponsor, and made it back alive—but I wouldn't be afraid to test this fortune again, because I believe you create your own fortune. I put in the effort with the proposals and the solution appeared from directions I wasn't even looking in. Just take my association with Mistral, Sandringham Yacht Club and the *Herald Sun*.

I'm not saying anything new here. We all know this stuff. We see miracles in movies all the time, yet often we don't believe that they actually happen in real life.

I now believe they do! I have experienced the power of the human spirit and it's changed me forever, and I'm thirsty for more.

So, what does the future hold for me? I'm eighteen, I have no real job, I haven't finished school and I think I'm allergic to suits and ties (either that or I'm colour blind).

But I have the most important ability in the world—the ability to dream. And no-one can take that away from me. For the last three months of the voyage I spent every moment I could in front of the computer documenting and putting into words my vision for the next trip. That proposal was finished before I got home and I'm now working on it.

So, what is it? Beau and I with three or four others will set out on an old-style 46-foot Polynesian catamaran built of timber with two gaff-rigged sails. We'll travel around the world again, but this time over three to four years, stopping off at the most remote and exotic places, such as the Amazon,

Beyond the Waves

Madagascar, Galapagos, the Spice Islands and Papua New Guinea.

We'll take only lanterns for light and a sextant for navigating. Electricity will only be used to charge the equipment for filming the voyage. That footage will be turned into a fifteen-part series on the themes of youth, environment, culture and adventure.

Now to the most important question: what have I learnt from sailing solo, unassisted and non-stop around the world?

I have discovered that we mustn't limit other people's abilities by our own. We need to encourage and help those around us, particularly our youth, with whatever their dreams may be, and then we'll start to see great things happen. I was just a normal kid with a dream who was serious about what I wanted to do. But without the support of my family, I would never have made it, and would have eventually lost enthusiasm with age and become like so many others—an unsatisfied grown-up who doesn't believe in himself.

There are many people out there dreaming of great things, and it's a good chance that your son, daughter, brother, sister or friend is one of them.

Believe and encourage them so they won't lose one of humanity's most prized assets—the ability to dream.

Appendix 1

Equipment List

Navigation
admiralty list of radio signals (4 volumes)
binoculars
parallel rulers (2)
sight reduction tables (4 books)
steering compasses (2)
barometer
divider
sextant
star finder

Safety
50 metres nylon line for drouge
four-man liferaft with double
    layered floor
safety harnesses (3)
drouges (2)
sea anchor
fire extinguishers (2)
harness lanyards (2)
medical kit including
    injections and pain
    relievers

Sails
#3 jibs (2)
mainsails (3)
spinnaker poles (2)
genoas (2)
spinnaker (1)
storm jibs (2)

Sail repair kit
hanks
leather
needles
heavy duty sail cloth
metal rings
palm

Equipment List

scissors
twine

sticky back sail cloth
webbing

Rigging
coil of rigging wire
Norseman fitting (various kinds)
spare halyards (2)
spare shackles (various sizes)
whisker pole
wire cutters

flemming wind vane
plough anchor and chain
spare heavy duty blocks (6)
spare vanes (7 sizes)
winch handles (6)

Tools
drill sets (3)
hammers
sockets
vice (fixed to bench)

hacksaw
hand drill
spanners
WD40 grease

Electronics
video cameras with underwater
    housings (2)
CARD radar detector
digital temperature gauge
emergency whip antenna
Hf radios (2)
mobile spotlight
radar reflector
Raytheon GPS (installed)
satellite e-mail equipment (Inmarsat C)
spare battery chargers and film
    and batteries

CDs (60, various artists)
autohelm tillerpilots st2000 (2)
CD stereo system with
    speakers
Garmin 48 handheld GPSs (2)
laptop computers and
    chargers (3)
Raytheon 24nm radar
Raytheon wind instruments
satellite phone
still camera
VHF radio (installed)

Electricity
AA batteries (50)
80-watt Solarex solar panels (3)
Batman digital battery readout
spare generator blades (3)

480 amp hour battery bank
air marine wind generators (2)
D-size batteries (20)

Electrical repair kit
acid

connectors

Lionheart

electrical tape
fuses
pliers
solder
wire brushes (2)

fluorescent globes
globes (various sizes)
Senson grease
spare light units
wire

Emergency grab bag
406 Mhz EPIRB
flares (rocket, orange, red, white)
handheld VHF radio
strobe light

breakable neon lights
Garmin 48 handheld GPS
PUR watermaker 1000
survival suit

Clothes
boots (2 pairs)
polar fleece jackets (3)
thermal underwear (3 sets)
tracksuits (6)
t-shirts (6)

hats (8)
sleeping bags (2)
wet weather gear (4 sets)
socks and jocks (8 sets)

Toiletries
baby wipes
cakes of soap (12)
toothbrushes (8)

bottles of shampoo (6)
toilet paper
tubes of toothpaste (10)

Miscellaneous
alarm clock
books (over 100)
pencils (box)
cooking utensils
erasers (6)
guitar in case
large tupperware containers (8)
multi-vitamins
pencil sharpeners (4)
small tupperware containers (6)
exercise books (8, various sizes)
waterproof Pelican cases (8)

beanies (3)
biros (box)
buckets (6)
dolphin torches (8)
fishing line (500lb)
jerry cans (12)
methylated spirits (200 litres)
pair of gloves (2)
school books and CD Rom
spare ropes
water (450 litres)

APPENDIX 2

Glossary

Aft	Towards the back of a boat.
Bilge	Lowest point of the hull's interior.
Boom	The horizontal pole that extends from the mast, holding the bottom edge of the sail.
Broach	To surf down a wave and turn side-on.
Bow	The front of a boat.
Clew	The lower aft corner of the sail.
Cockpit	The area where a boat is steered.
Companionway	Stairway and opening leading from the cockpit to the cabin.
Deck	The platform running the length and width of a boat.
Drogue	A small webbed canvas parachute about one metre in diameter towed behind a boat to slow it down in rough weather.
Fore	Front section of a boat.
Foredeck	The forward part of the deck.
Forepeak	The front section of the cabin, where the bow rises up. Used for storage on *Lionheart*.
Forestay	The wire from the top of the mast to the bow.
Furler	Mechanism to pull genoa in and out.

Lionheart

Genoa	Largest front sail, attached to the forestay.
GPS	Global Positioning System—the navigation tool enabling accurate latitude and longitude readings for location.
Gybe	To catch a tail wind on the opposite side of the mainsail, swinging the boom across a boat.
Halyard	The rope used to lower or raise a sail via a pulley system.
Hanks	The small rings or clips that attach the sail to the stay.
Hove to	To point the bow into the waves, using a small amount of sail, to stop its progress.
Knot	Measure of nautical speed: 1 knot equals 1 nautical mile per hour.
Lanyard	A short rope to secure a safety harness to a boat.
Lee side	The side of a boat or land protected from wind and rain.
Leeward	The direction to which the wind blows.
Lifelines	Small safety fence around the perimeter of the deck.
Luffing up	To point a boat into the wind.
Lying ahull	The process of pulling all sails down to ride out a storm.
Mainsail	The principal sail of a boat, attached to the mast.
Mainsheet	Rope controlling the angle of the mainsail.
Mast	The vertical pole holding the sails.
Meridian	A line of longitude.
Nautical mile	The measure of nautical distance: 1 nautical mile equals 1.852 kilometres.
Outhaul	A rope used to pull a sail along the boom.
Pitch-pole	To tip a boat end on end, usually by sailing down a wave and nose-diving into the wave ahead.
Port	To the left of a boat.

Glossary

Reef	Reduce the amount of sail area to slow a boat or reduce strain on rigging.
Rig	The mast, sails and wires on a boat.
Sea cocks	Water inlets in the hull of a boat.
Shrouds	The wires from the mast to the side of a boat.
Spreader	Horizontal bar on the mast which wire runs through to support the mast.
Starboard	To the right of a boat.
Stanchion	The small posts supporting the life lines around a boat.
Stern	The back end of a boat.
Washboard	Removable boards that seal the companionway.
Wind vane	Self-steering device mounted on stern that uses wind to steer.
Windward	The direction from which the wind blows.

APPENDIX 3

Parts of the Boat

1	baby forestay	14	furler	31	pull-pit
2	backstay	15	galley	32	radar
3	books	16	genoa	33	radar detector
4	boom	17	head (toilet)	34	reefing lines
5	bow	18	inner forestay	35	rudder
6	bunk	19	jerry can storage	36	running backstay
7	cockpit	20	keel	37	solar panel
8	collision bulkhead filled with foam	21	lazarette	38	spare sail storage
		22	mainsail	39	starboard side
9	companionway slide	23	mainsheet	40	stern
		24	mast	41	storage
10	companionway steps	25	metho tank	42	switchboard
		26	motor	43	tiller
11	forepeak food storage	27	navigation table	44	tricolour light
		28	portholes	45	washboards
12	forestay	29	port side	46	water tank
13	forward hatch	30	propeller	47	wind vane

Parts of the Boat

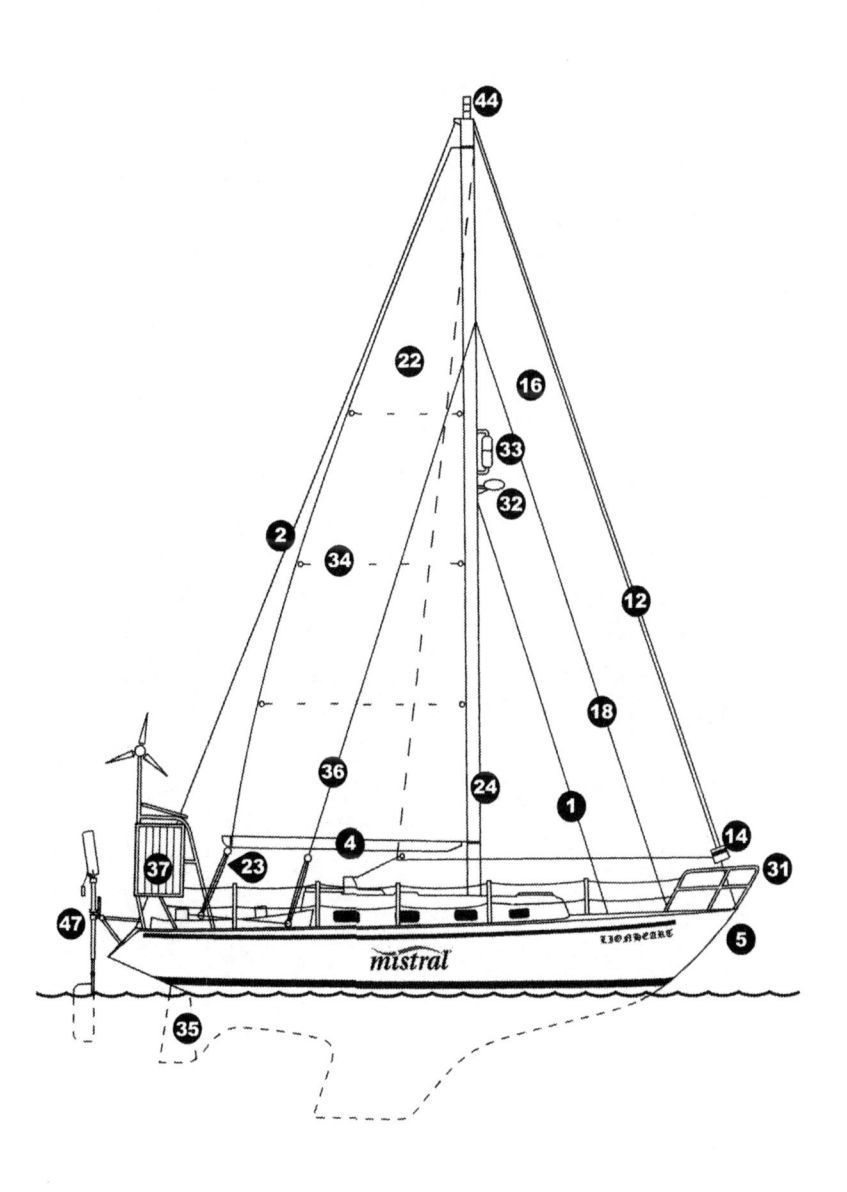

Lionheart

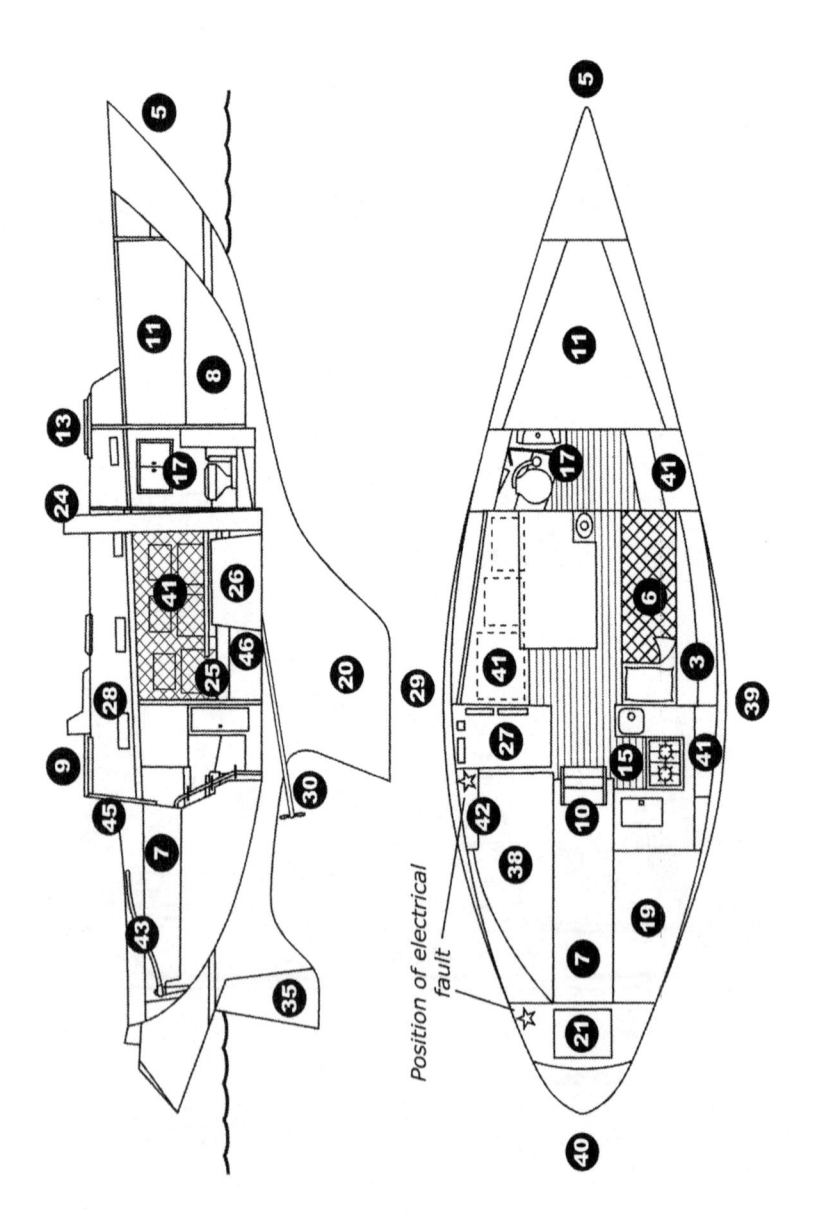

Position of electrical fault

Acknowledgements

A special thanks go to all these people who, in some way, put considerable effort into my life and the publication of this book.

First of all, John Hill, for his passion at helping out wherever possible and his eccentric ideas, which often turn out to be genius.

Phil Carr, for his huge contribution in time for getting the boat ready at the last minute.

Scott Eccleston and Richard Hewett from Sandringham Yacht Club for the organisation of dinners and the day of my return. Kevin Wood, the commodore of the yacht club, and Stewy Howarth, the radio officer.

All the operators at Sydney Radio who took my calls and chatted briefly with me.

Dr Broomhall, who made several visits to our house on Tuesday nights to advise me on medical procedures and on what to take.

Jacinta Oxford, for an unbelievable effort beyond the call of her job in regard to the nutrition of food, menus, purchasing and packing. Much appreciated!

Roger Badham, who gave me peace of mind every day with his weather forecasts for the whole journey. Thanks, it was a huge job!

Barbara Pesel, for organising the media and synchronising everything—a rather large job and well done.

Don and Margie McIntyre, for not only subsidised equipment but positive mental support before I'd even left.

Bob Charles, for assistance in various areas.

Andrew Burley, for nothing in particular but just constant energy and ideas and a sounding board for theories. Thanks!

Lionheart

Matthew Gerard, for his big heart, which to this day I can't comprehend.

Steve O'Sullivan, who seemed to be the pivot on which all the great assistance came. Without that help I'm sure things would have turned out differently!

Sue Hines, the publisher, for a quick reaction and foresight in signing on the book. Mark Davis, for the cover and art layout and Foong Ling Kong for the editing.

Phil Gregory, who pushes me and excites me with his passion for the work we plan to do, and for getting all the smaller details correct that I didn't even know existed.

Ed Gannon, who has seen the trip from the start and wrote this book with me. We were both beginners as far as writing books go and I thank him for helping me express myself in an understandable fashion and for all the extra work and long hours he squeezed in. Thank you heaps!

And finally, Mum and Dad. What can I say other than thank you so much for letting go, probably the hardest thing, but the best choice you could have made.

Sponsors of the journey

Abbott Australasia
Australian Milk Marketing
Autostream Propellers
Codan
Dandenong Ranges Tourism
Distance Education Centre of Victoria
Gelb & Ischiaa Pharmacy
GPM Systems
Herald Sun
Hewlett Packard
Kodak Australia
Le Tan
Mistral
McIntyre Marine
Musashi
Parkin Electrical
Sceneys
Senson
Sanitarium
Solarex
Solid Solutions
Strahmar
Snowgum
Unilever
Visual Entertainment
Wesley College